Random Tales of a College Math Professor, Over-Easy, on Wry

MICK NORTON

ISBN:146811638X
ISBN-13:9781468116380

DEDICATION

The first statistics course I ever took was taught by Duane Norman. After the course was over, he got me my first paid consulting gig – analyzing some data for the Kirksville, Missouri public schools. One of my vivid memories is of his opining in class one day that dedications in books – to the spouse, or mother, or whatever – were mushy and misguided. "If I ever write a book," he said, "I'm dedicating it to myself. After all, I'm the one who's doing all the work." Well, he didn't write that book. But he did lay two bricks in a foundation. And for that, many thanks.

The other dedicatee also was a teacher, and a darned good one. Jeanette Bickley helped shape, mathematically, many individuals over the course of her career at Webster Groves HS and beyond. The mathematics teaching community couldn't have had a better role model. And for that, many thanks.

CONTENTS

ACKNOWLEDGMENTS

A friend – a non-mathematician – suggested that I think about writing a book of math stories. Saying that she wanted a wry humor to come through, she encouraged me to write the stories like I tell them. And so, my deepest thanks go to Marcia White. Because one goal was that there be no math prerequisite for reading the book, Marcia also provided a caution. Accordingly, when describing the mathematics or statistics connections in these stories, the author has tried hard to not be pedantic. Additionally, he has tried hard to avoid using words like *pedantic*.

1 EVEN THE POLICE NEED TO SOLVE MATH PROBLEMS

Mick was eating with the usual group of math professors who ate lunch in the faculty lounge. But he was not involved in the discussion as much as usual. On this particular day, the lunch banter didn't capture his full focus. A policeman was to meet him in his office at 2 pm. Department chair Bill Golightly had asked him if he would like to do some community service. The City of Charleston Police Department needed help with a math problem. Wondering what kind of math problem a police department would need to solve, the professor told Bill he'd be happy to help. The 2 pm meeting was to be with a Lt. Robert Roberts.

At 2 on the dot, the lieutenant, accompanied by another detective, knocked on the open office door. Roberts explained that they were trying to decide whether a woman's death was a homicide or a suicide. She'd had a boyfriend who was really bad news. Of course, one can never predict for sure the amount of bliss and joy might accrue from a relationship with someone who possesses a long police record and a history of forcing people to play a version of Russian roulette with him. She had recently made the decision to leave him for a sailor, but before that could happen she wound up dead.

"I can show you what I think happened," offered the lieutenant. "But first I need to take the bullets out of this."[i]

He produced a revolver and made an exaggerated display of taking the bullets out of and of showing the professor that the cylinder was empty. There probably was some standard operating procedure that required such a demonstration, particularly if pointing the gun at someone was to be part of

1

the demonstration. Just as the display was in progress, three of Mick's students stuck their noses inside the open doorway. It was, after all, office hours. Their faces registered concern. *Hey, guys, doesn't this happen all the time during my office hours?* popped into the professor's head.

"Hey, guys, could you come back in about 15 minutes? I'm kind of into something right now." They nodded and left, looking somewhat reassured but still apprehensive.

After repeating that they hadn't ruled out suicide yet, Roberts moved about the room as he explained and showed how he thought the boyfriend might have shot the woman. She was found lying on a bed. The bullet had entered the right side of her head and exited on the left, coming to rest on a pillow. She was highly intoxicated when she died, and there were powder burns on both of her hands. She was left-handed. Using both hands, the lefty might have held the gun to the right side of her head and pushed the trigger with her thumb. They wondered if it was possible for her to have committed suicide and fallen back to her final position, given where the bullet landed. What they had was a geometry problem.

"Can you tell from looking at these whether you can help us?" asked Roberts. The other detective produced an album of crime scene photos, which the professor opened and began to study. Roberts started to expand on his explanation about what he thought had happened, but Mick put a hand up as a request for quiet. He was focused on the photos, looking for any math openings they might provide. She was lying face-up on the left side of the bed, her head a couple of feet from the headboard. Her posture was that of a Visible Woman anatomy toy. On the floor, below her right hand, was a handgun. The only pillow on the bed was near the headboard on the right side of the bed. The bullet, clearly visible on the pillow, had left a crease and skid mark, or maybe a scorch mark, on the pillowcase. The photos showed that the skid mark was about two to three inches long. He settled on one photo that did the best job of showcasing the pillow and the headboard end of the mattress.

"I can't give you the path of the bullet," offered the professor. "But I can give you the vertical plane that contains the path."

"Can you explain that?" Roberts was most attentive.

"Well, over such a short distance, we can assume the bullet traveled in a straight line from revolver to pillow. Unless … would passing through her head change the path of the bullet?"

"No."

2

"OK. So next, the bullet hits the pillow. Now, hitting the top, flat surface of the pillow isn't going to make the bullet veer to the left or right. Basically, the bullet would stay in the same vertical plane. That means the skid mark lies in the same vertical plane as the path of the bullet." Mick used hand sweeps to help illustrate the plane. Miming with a different set of hand motions, he explained further. "The bullet could come at the beginning point of the skid mark from a shallow angle or from a steeper angle. But whatever the angle is, all possible bullet paths would lie in that same vertical plane."

"Would you be free Saturday morning to help us set up a model?"

"Sure."

And so on Saturday, Mick reported to one of the rooms at the Charleston Inn on Lockwood Drive. The motel room was generic. There was absolutely nothing special about it. On the scene was a mix of uniformed policemen and plain-clothes detectives. One of the officers was a woman the approximate size of the victim. Lt. Roberts made all the introductions.

The professor was impressed with what was on hand for the model construction. There was a Styrofoam head, the kind someone might store a wig on. The head had a metal rod passing through it so as to match entry and exit wounds. The entry point appeared to be near where a right Styrofoam ear would have been, if the head had had ears, and the exit point on the other side was a little closer to the face. Also present was a vertical post secured to a base, the whole thing made out of two-by-fours. It was something like a cat's scratching post, only much taller and without the carpet. They also had brought tape, string, and whatever else was the thing to bring to crime scene reconstructions. There was one pillow on the bed, positioned as shown in the crime scene photos.

Earlier, using that one crime scene photo and a protractor, Mick had measured the angle formed by two key lines: the edge of the mattress that butted against the headboard and the extension of the skid mark. Sliding the protractor along the edge of the mattress while keeping string that was stretched across the protractor taut and at the correct angle, Mick found the position that was needed to make the string pass across where the skid mark on the pillow would have been. The approximate beginning and end of the skid were then marked on the pillow. With the protractor in the same spot and the professor certifying the angle, an officer held taut a length of string that was long enough to stretch over the skid mark and beyond to the left side of the mattress. The officer positioned the vertical post so that the string just grazed the post. The vertical line that was that edge of the post and the beginning point of the skid mark identified the approximate location of the plane that contained the path of the bullet.

"Is it OK to replace the metal rod with a long piece of string?" queried the professor.

One of the detectives took one end of the long piece of string, tied it to one end of the rod and, holding the other end of the rod, pulled the rod out of the head, thereby threading the string through the head. Anticipating what needed to happen next, the officer taped one end of the string onto the pillow at the beginning point of the "skid mark."

Holding the string taut against a point on the post would identify one of a potential infinity of possible bullet paths. For any given path under consideration, the head could be positioned by sliding it along the string and/or by rotating the head using the string as the axis of revolution. Any such position of the Styrofoam head would be a candidate for the position the woman's head was in when the bullet passed through it.

It then became a matter of trying a variety of possible bullet paths and head positions on those paths, determining for each such position whether the policewoman could sit on the bed, position her head to match (with her body tagging along), have the upper half of her body fall backward, and match the final position of the dead woman.

The first path they tried was one that would make the Styrofoam face be vertical. Because there was a slight tilt to the path the bullet took through the head, the string had to be so low on the post that the woman's body would have to have been imbedded in the mattress. One down, infinitely many to go. After a bit of trial and error, a narrow window of places on the post that the string could touch was established. Very narrow. Further, positioning the head on any bullet path in that window forced the policewoman to be in the middle of a sit-up.

"Is there any other way the body could have wound up where it did?" Roberts asked.

Mick held the string against the post in a place that gave a plausible bullet path and pointed toward the top of the door to the room.

"If you extend this line in the direction of the door, in theory, her head could have been anywhere on the line. Say, starting from the door, she could have run toward the bed, jumped up in the air doing a Fosbury flop, shot herself while she was in the air, and landed on the bed."

4

"That's ridiculous." Of course Roberts, who always came across as serious, was right. But also of course, he had asked a math question and the professor had provided the answer.

"Yes, exactly so. The only possibility that is realistic is that she was in the middle of a sit-up when the bullet went through her head."

Roberts studied the head and taut string for a moment and began maneuvering the head on the string-path, twisting the orientation a little so that the face would have been squared up with respect to where the shoulders would have been for a body that was mid-way into a normal sit-up.

"You're bending the path of the bullet," cautioned Mick as he pointed with his free hand, drawing attention to the two kinks in the string.

"You mean her head was tilted slightly to her left?"

"Yeah. Maybe she was trying to turn her head away from the gun."

"Well, that pretty much does it, then," announced the lieutenant. The pronouncement was made with a "Book him, Dano" finality.

Roberts was now convinced that it was a homicide. How many dead-drunk, left-handed women would commit suicide by holding a gun against the right side of the head with both hands and push the trigger, while being half-way up into a sit-up, looking slightly to the left? A body position like that was more consistent with her having been involved in a struggle with someone and trying to push the gun away at the instant the bullet went through her head.

Later, as the murder trial progressed, the professor followed the daily reports in the newspaper. He was told to be prepared to testify, but never had to. Lt. Roberts called one day to say that the crime scene reconstruction would not be allowed. He said the reason was technical – the Styrofoam head did not have nipples affixed at the entry and exit sites. Roberts spoke consolingly, assuring the professor that not allowing the reconstruction had nothing to do with the merits of the reconstruction. Although Mick's feelings were not hurt in the slightest, he did appreciate that Roberts felt there might be a need to be nurturing. All the way around, there was a lot to respect about the lieutenant's professionalism. Also, the professor looked at the events of the last few days as a glass half full. Although the reconstruction was not allowed in the trial, good had come from it. The police were convinced that they needed to pursue a murder investigation.

Mick actually was not unhappy at not being called as a witness. Testifying on statistical matters, either in court in an age, race or sex discrimination suit, or before a university honor board in a case where a student had been accused of copying off of another student on a multiple-choice exam, was one thing. In a murder trial though, he didn't want the bad guy to get off and later remember the math guy, the witness for the prosecution who talked about vertical planes. The police had amassed other evidence, and the prosecuting attorney had discredited the defense witnesses whose courtroom testimony, which provided alibis for the accused, differed from earlier statements they had made to police. The former boyfriend was convicted.

2 HURRICANES AND BOWLING BALLS

A left-handed bowler's hook to the right is intended to come at the headpin from the "Brooklyn" side. If the ball somehow doesn't hook, it can wind up in the left gutter. If the hook bites way too early, the ball will go into the right gutter. Either of these is good when the pins are looking for a little peace.

During Hurricane Season, the West Coast of Africa spits a random number of tropical storms out into the Atlantic Ocean. Some of these become hurricanes that can ravage the East Coast of the U. S. As the storms course westerly across the Atlantic, a good number of them hook northward. To those who live on the East Coast, having the eye of a hurricane make landfall thirty miles north of where you live means, probably, high winds, hard downpours, policing up some branches from your yard, and not having to fix anything on your house. But if the eye makes landfall thirty miles south of your house, it's bad news. Just ask the residents of McClellanville about 1989, the year when the eye of Hurricane Hugo went directly over Charleston. What happened to Charleston was bad – very bad. And the well-known city got appropriate play in news reports. But the national news coverage of poor little McClellanville, the 6-pin, suffered from name recognition.

```
10
      6
  9       3
      5       1
  8       2
      4
  7
```

Ah but we get ahead of ourselves. People who live on the East Coast would number pins in different ways depending on where they live. Mick, a Charleston resident, viewed Charleston as the 3-pin. Also to his way of thinking, Beaufort, to the south of Charleston, was the 2-pin, and midway between Beaufort and Charleston – that is, about thirty miles south of Charleston – was the headpin, or 1-pin. One common way of knocking down the 3-pin would be to have the eye – oops, ball – hit it head-on, such as Hugo did. The other most common way would be for the ball to have a trajectory across both the 1-pin and the 3-pin, Brooklyn-style.

Sitting at the southernmost edge of the array of pins was the Florida Keys, all of which constituted Mick's 7-pin. Many balls that didn't hook would miss the 7-pin – and all the other pins – and go harmlessly into the left gutter – the Caribbean. Of course, the people who live in the Caribbean would have a different numbering system.

And that part of North Carolina from Wilmington to Cape Hatteras, land that juts out prominently from the rest of the East Coast, was the 10-pin. Poor Wilmington, et. al. For any bowler whose hook to the right is unpredictable, there are lots of trajectories that will carry the 10-pin. Even if the ball doesn't hit any other pins, it might take out the 10-pin just before falling into the right gutter – the mid-Atlantic.

Hurricane season is a time when probability comes into play. A natural mathematical model that statisticians use and which does a good job of approximating some of the hurricane-related random numbers that Mother Nature generates, such as the amount of time that elapses between consecutive storms, is called a *Poisson process with nonstationary mean*. Never fear, reader, there won't be a test on this. At the beginning of each hurricane season, meteorologists study weather patterns and predict the number of tropical storms there will be. Suppose these experts say to expect fifteen tropical storms this season. Stat folks can then use this information and the

Poisson model to answer arcane questions they can put on hour exams in their introductory stat classes. For example, if 15 is how many to expect, what is the probability that there will be exactly 15 (10% chance), or exactly 17 (8.5% chance), or at least 17 (33.6% chance), and so on.

Further, suppose that forty percent of tropical storms reach "category 4" hurricane status. Then to expect 15 storms is to expect six "category 4" hurricanes, which follows by direct application of one of the formal mathematical "rules of expected value," the existence of which illustrates that every once in a while, mathematical theory agrees with common sense.

Hurricane season is also a time for guilt feelings. When a hurricane is headed toward Charleston and a hook either hasn't started yet or has started but hasn't become pronounced, Charlestonians begin to think: *let the hook really bite, so it hits North Carolina.* It is a positive reflection on their character that Charlestonians feel absolutely guilty for thinking that. It is likewise a reflection on their great character that they are most understanding when coastal residents of Georgia and north Florida root for oncoming hurricanes to hook northward and make landfall in South Carolina. Wishing for a self-serving hook to kick in – or not kick in if the thing is headed safely to your south – has exactly the same amount of influence as wishing that next year's weather patterns keep the west coast of Africa from bowling many frames.

3 HAVE STATISTICS, WILL TRAVEL

Academics didn't make a barrel of money. Everybody knew that. It was the intangible highs that made it a great profession, such as helping students have light bulb moments about mathematics. The satisfaction that came from watching students blossom was hard to top. And then, of course, there was performing the craft itself. Solving math problems as part of the job requirement was like getting paid to have a hobby. There was a recreational aspect to it, a little like solving Sudoku puzzles rated *beyond-evil*. Investigating alluring math conjectures that he or others had posed, finding proofs of the conjectures when he could, having the proofs accepted by peer referees, and seeing the results in print as journal articles produced incredible highs. And as with drug highs, there was an addictive aspect. It was easy to get used to having your creative moments lead to seeing your name in print, and disconcerting to go through a slump.

That the Sudoku analogy was weak is an understatement. Solving any particular Sudoku puzzle had no benefit to the world in general, whereas solving previously unsolved math problems or proving your own conjectures added to the known knowledge base of mathematics. Also, the solutions to many math problems resulted in immediate real-world applications.

Between grading, committee work, making lesson plans, teaching, professional development and doing research, a professor typically would work 70 or more hours a week during the school year. But except for meeting his classes at their regularly scheduled times, he had the flexibility to manage the 70 hours however he wanted. If he needed to see the dentist, any time or day of the week would be OK, except for Mondays, Wednesdays and

Fridays from 9 to 11 am and Tuesdays and Thursdays from 12:15 to 1:30 pm. One colleague in the department liked late afternoon classes and late office hours. He'd arrive at the department around 2 pm, meet his classes, and work in his office until around 4 am. He liked quiet. His work break would be a workout out at the gym at 1 am.[ii]

Even though doing all the things he was supposed to be doing could more than fill the time, the itch to somehow spice up his professional life developed after about five years on the job. Something that would help scratch the itch was consulting. His first paying gig as an expert witness in statistics was about to materialize, the result of a random event. A medical student at the Medical University of South Carolina (MUSC) was accused of cheating on each of four multiple-choice exams that were given in a course. Let's refer to the student as Mr. Romanize. A number of students had come forward to make allegations that Romanize had copied (possibly) off of: four students on exam 1, three students on exam 2, three students on exam 3 and four students on exam 4. Fearing that appearing before the MUSC Honor Board could cost him his future medical career, Romanize hired an attorney. The attorney, now in need of somebody who could do a statistical analysis, knew someone who knew Mick. And one thing led to another, as they say.

One afternoon, the professor arrived at his first meeting with Romanize and the attorney at the law offices of the latter. Mick was looking forward to meeting the med student. Anyone who could galvanize not just one, but a group of fellow students to come forward with allegations that he had cheated, not just on one, but on four exams, had to be an interesting fellow. First impressions, though, were that Romanize didn't exude guilt, and he didn't exude innocence. Nothing about him stood out at all. During the let's-get-acquainted part of the meeting, the student offered that he had no idea what had possessed the students to come forward, that no one disliked him that he was aware of, and that he had not cheated on any exams. After trying to take measure of the student's demeanor, the professor had no idea which one of two extremes was the more likely. Either Romanize was absolutely innocent or he was a good candidate for a job in espionage or as a secret agent.

The attorney explained that a statistician at MUSC had compared the answer sheets of the students and done a statistical analysis that was damaging to the student's case. The analysis and raw data would be made available to Mick if he could help. The professor said that he would like to give the data and analysis a look, and then would let him know immediately whether he

11

could help. Mick also proposed an hourly fee, which would kick in once he'd decided that he could help.

"I'd really rather not use statistics at all," said the attorney. "When a scientist or mathematician says something, people are too prone to think that something has been proved, not just beyond a reasonable doubt, but proved in the mathematical sense. But I don't have much choice here. MUSC is using statistics, so I have to.

"One of the non-statistical things we had been talking about before you came was that we need to explain how two students could agree on an unusually large number of answers when no cheating took place."

Romanize chimed in. "When people study together and use the same study materials, they should be expected to agree a lot on answers, even on wrong answers. And I did study with several of those students. There is also just plain, bad luck."

Mick was not on anybody's dime yet, and given what the other two were about to resume discussing, he recognized that it was an appropriate time to leave. But first he felt obligated to inject a statistical thought into their discussion.

"Sometimes, there are unknown explanations lurking in the background, explanations beyond chance, that can cause two things to correlate – things that might never occur to you to consider. In statistics, there is even a term for these unknown things. They are called *lurking variables*."

"Can you give us some examples?" queried the attorney.

"That would be something for you guys to brainstorm about." The professor turned to Romanize and continued. "You're the one who knows these other students and how you interact with them. Think about your routine, and about your interactions with them, and see if anything clicks that might explain why you would agree a lot with their answers. Be sensitive to whether there might be explanations outside the box. You might not come up with anything, but it's worth thinking about." Mick started to get up.

The attorney looked interested and puzzled. "An example would really help."

"Well, you're right. I didn't give you much guidance, did I? Well, off the cuff … the first thing that comes to mind is a little far out."

"Go ahead anyway."

"OK. When I was in high school, I remember watching an episode of Rod Serling's *Twilight Zone*. The setting was the old West. In a small town on the prairie, one of the residents had been murdered. Next to the body, there

12

was a wallet with the initials MFQ, a rare combination of initials. No person in town had them. The deceased was a good guy with no known enemies. Everybody in the town knew everybody else, and they were confident no one in town had done it. But there did happen to be one stranger who was passing through town at the time. His name was Michael F. Quentin, or something like that. There was nothing to tie the stranger to the victim, and nothing to indicate that he had ever been near the vicinity where the body was found. Based on the unusual set of initials, a jury found the guy guilty and he was hanged.

"Afterward, some of the townspeople were having second thoughts about having hanged the guy. But what other explanation could there have been for the wallet?

"Not long thereafter, the general store received a large shipment of leather goods, purses, saddles, and so on. Every one of them had the initials MFQ stamped on it. They were made by the Mighty Fine Quality Company."

After Mick had delivered the light but corny illustration, the other two nodded, indicating that they got it. Mick made his farewells and headed back to campus. He chided himself on the way, feeling guilty for not having come up with a classier example, one that was more relevant to the given situation. *The Twilight Zone? Really, now. Couldn't I have done better than that? In my defense, though, it is Romanize who has the most background information.*

Several days later, the professor had read the statistical analysis of the data in the Romanize matter. Boyd Loadholt, a statistician who worked at MUSC in the Department of Biometry, had compared Romanize's answer sheet to the answer sheets of each of 14 students who had been named in the allegations. In some cases, Romanize's answer sheet was not particularly similar. But on each of the four exams, there would be one or more students whose answers agreed significantly with Romanize's.

Mick thought the analysis was pretty slick. In comparing the answer sheet of Romanize to that of another student on a given exam, each exam question contributed a numerical input toward a number that Loadholt called an "index of agreement." All of the inputs, one per question, would be added together to produce the index of agreement for the two students – the greater the sum, the greater the agreement between Romanize and the student. Each question's input toward the sum was number between zero and one. If the two students gave different answers to a question, which indicated no copying on that question, the input for that question was zero. If the two students gave the same answer (it made no difference if the answer was correct or incorrect), the input for that question would depend on the percentage of all the students who answered that question in the same way. The more rarely that response was given, the greater the input would be. Checking extremes, Mick noted that if everyone in the class gave that same answer, the input would be zero, whereas if only Romanize and the other student gave that answer, the input would be very close to one. So the way to have a large index of agreement with another student was to match that student's answers a lot, particularly rarely given answers.

Suppose Student A was one the students that Romanize had been accused of copying from on an exam, and that the index of agreement for those two students had been computed. To know whether the result of the computation was large enough for a statistician to deem *significant*, it would be necessary to also compute the index of agreement between Romanize and each student who took the exam. If, say, 150 students took the exam, then altogether there would be 149 indexes to compute. The most damaging thing for Romanize would be for his index of agreement with Student A to be the largest of all of 149 indices. Interestingly, that very thing had happened on each of the four exams. So the statistical case against Romanize was a good one.

The only criticism that Mick had with the analysis was how probabilities had been computed. For the example in which Romanize and Student A had the highest of all 149 indexes, the professor would have assigned a probability of 1/149. One could then say, *We were told that you may have copied off of Student A. Besides you, there were 149 students who took the exam, and we found that your answer sheet agreed more with Student A's answer sheet than it agreed with anybody else's answer sheet. If no copying occurred, the probability that you would agree more with A than with anybody else is 1/149.*

Loadholt's probabilities were tied to normal distributions, and so would be slightly different than Mick's. The professor understood what Loadholt was doing and why. The issue was more a matter of style than of substance.

Not long thereafter, the professor had another meeting with Romanize and his attorney.

"I've looked over the statistical analysis done by MUSC," began the professor. "I can take the case. Dr. Loadholt did a pretty good job. There's not much to criticize. I can think of only one negative thing to say about how he puts together his index of agreement. If somebody's goal is to *create* the largest possible index of agreement, he should copy off of the worst student in the class, or someone who gives rarely given answers, or someone who just guesses on every question. But the other side could easily respond by asking, *Why would anybody who wants a good grade do that?* But then you could say, *If he had a need to copy from someone, maybe that means he does not make good decisions in general, such as when choosing a copyee.* So a criticism of the index is there if you want to use it.

"The materials I was given also show that the M.D. from the Department of Pediatrics who is chairing the hearing did his own analysis that combines the four exams into one package. It's a natural thing to do. So I would like to do my own computation – a single probability for the extent of agreement that occurred on all four exams, under the assumption that no copying occurred. It's something that ought to be looked at anyway, and the other side did it. I'm not sure, though, that the way he packages everything together is the way it should be done. I need to think that through some more.

"Finally, you should know that I would have computed probabilities a little differently than Loadholt did. You can't say that he was wrong in how he did it. It's mostly a matter of personal style. You should know, though, that the probabilities I come up with when combining all four exams could turn out to be more damaging to your client than the other side's. Now that I've said that, do you still want me to go ahead?"

"Sure. Actually, if that happens, it would be easy to put down the use of statistics altogether. If they say the chances that the exams would agree as much as they did with no copying going on are five in 100,000 and you say it's one in 100,000, that's really a good thing. That allows me to say that their side has someone who is supposed to know statistics, and my expert is supposed to know statistics, and the two don't agree. So what does it all mean? Using statistics muddies the waters by bringing up side issues that don't shed any light on whether copying occurred."

According to MUSC rules that were in place at the time, the student's lawyer could be present, but the professor was not allowed to appear. His report, however, was presented. Several days after the hearing, Mick was at MUSC's Department of Biometry, meeting with the department head on a matter unrelated to Romanize. When the meeting was over, he stopped by Boyd Loadholt's office and knocked on the open door. Boyd looked up from his desk.

"Hello, Mick Norton."

"Hey, Boyd." Both were smiling, a recognition that they had been involved in a contest of sorts, and that it was over. Mick sensed a familiar feeling. This was the same kind of camaraderie he had experienced when he had wrestled in high school – on those occasions when he'd get to talk with his opponent from the other team about the match they'd wrestled. "The lawyer was going to call me and let me know whether the student was found guilty, but he hasn't called yet. Do you know the outcome?"

"Sure. They didn't kick him out of school, but they did make him take the course over."

"Actually, he probably should consider himself lucky with that. The statistics were pretty incriminating."

"Yes, any way you look at it. I read your report, and knew immediately what you had done. I was there for the entire hearing. The panel did its best to be fair, and they can feel pretty good about what they did. There is only one thing I didn't care for. The student's attorney. My God. He brought up the dopiest story you could imagine about some cowboy in the old West who was hanged because his initials were on a wallet. It was God-awful. Be glad you weren't there. Where on Earth could he have gotten something like that? And we all had to sit there and listen to it."

4 YOU NEVER KNOW WHAT YOU'RE GOING TO GET

His phone was ringing. It was Leomy, the math department's office manager.

"Could you help this person?" she queried.

"Aha. Somebody in Charleston has a math question. Sure, put them on."

Sometimes, callers would have interesting questions. Sometimes not. Sometimes, the question could be answered immediately. If, however, the problem could be solved quickly but not off the cuff, he'd take a phone number and promise to call them back in a few minutes. Responding to calls like these was good PR for the College and good for Charleston and the surrounding region. Once in a while, someone would call with a problem that would take so much time to solve, it would have to be viewed as a consulting project. In this case, if the professor had the time and the problem was interesting, he would talk about charging an hourly rate – which sometimes ended the call. Sometimes, Mick would refer the caller to someone who was better suited to answer the question. And sometimes, if the caller was in no rush for a solution, the problem might become a worthwhile project for a student. Listening to a caller's math question was like taking a piece of candy out of Forrest Gump's box of chocolates.

"Hi. This is Mick Norton. How can I help you?"

"Hello. This is Art Istenhausen. I have a math question. I'm drawing plans for a stone wall that's got to have a circular archway in it. The archway needs to be eight feet tall at its highest point and 30 feet across at ground level. I've been trying to draft a drawing, but I can't figure out what the radius of the circle is. I've tried different ones and they just don't work.

Actually, they somehow don't even come close. I need to have something by 4 o'clock. Can you help?"

"Sure. I'll have to call you back. A student with an appointment just this instant walked in the door. I know what he wants and it'll be quick. Can you give me your number? It should be no more than 10 to 15 minutes. OK?"

The professor drew the following sketch on a pad and then proceeded to help the student.

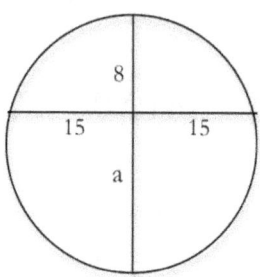

When the student left, Mick returned to the pad. Adding 8 to a would give the diameter of the circle. Since the 30-foot archway base was perpendicular to the diameter, there was a nice relationship between the four lengths pictured: $8 \times a = 15 \times 15$. From this it followed that $a = \dfrac{225}{8} = 28.125$ feet, so that the diameter, $a + 8$, had to be 36.125 feet. This meant that the radius had to be 18.0625 feet. Mick called the number Mr. Istenhausen had given him.

"This is Art Istenhausen."

"Hi. It's Mick Norton. Your radius is 18.0625 feet."

"No It can't be."

"Pardon?"

"That's too big. It won't work."

The guy sounded adamant. No doubt, 18.0625 was not near the realm of values that he thought were worth trying. The professor figured Istenhausen to be a draftsman or architect, someone who probably had drafting tools right there.

"Unless I've made an error somewhere, it ought to work. Try it and see."

"It can't work."

"Why not try it and see what happens?"

There was a pause and silence at Istenhausen's end. *The proof is in the pudding*, thought Mick. *He should come back happy.*

"OK, so how did you get that?" The caller sounded irritated.

The professor told Istenhausen to sketch a circle and explained how to draw the two lines inside the circle and label the four line segments. He then explained that the product of the lengths of the two horizontal segments had to equal the product of the lengths of the two vertical segments.

"Where did you come up with that?"

"It's not a result I use much, just something I remembered from way back in high school geometry."

"Is there a formula I can use that will give me the radius if I change the height and width of the doorway?" Again he sounded irritated.

"Sure. Let h denote the height, w the width, and let the radius be r." Mick instructed the caller to re-label the sketch as follows:

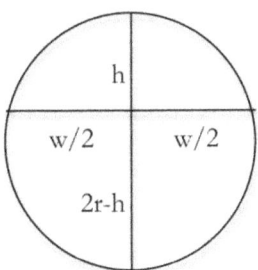

The professor went on. "Then $h \times (2r - h) = \dfrac{w}{2} \times \dfrac{w}{2}$, which means that $2r - h = \dfrac{w^2}{4h}$. Solving for r gives $r = \dfrac{1}{2}\left(h + \dfrac{w^2}{4h}\right)$. If you enter values for w and h into the right side, the formula gives you r."

"So if I want the height to be 8 and the ground-level width to be 20,…"

"Go ahead and try it. Tell me what you get and let's see if we agree," suggested the professor.

There were a few moments of silence.

"10.25."

"Yes, that's what I get, too." Mick then heard a click. Istenhausen had hung up.

19

"And you're welcome, I'm sure," said Mick as he hung up the receiver. *He sure sounded ticked off. Did I insult him when I said it was something I'd learned in high school? I certainly wasn't trying to. Why would anyone expect to remember a geometry fact like that unless they were interested in math? What was more likely was that now that he knew what the radius had to be, a real circular arc for a doorway didn't look right, which meant the guy would have to re-plan.*

5 BEING DIFFERENT AND BEING SIGNIFICANTLY DIFFERENT ARE DIFFERENT

The math professor was walking across campus on the way to his office, a bag of peanut M&Ms from a vending machine in hand. The imperative issue of the moment was whether to tear the corner of the bag open and begin eating enroute or to wait until he got back to the office. He decided to view the issue in terms of what would taste best. His conclusion was that yes, absolutely, he would savor the taste more if he weren't walking. Taste was the chosen criterion that day. Of course, for the question of whether to eat M&Ms en route, he always made taste the sole criterion, and so always waited to eat until he got back to the office. Mathematicians are known for reducing problems to ones they've solved before.

Just as he pulled the door open to exit the dominion of the hot afternoon sun, he heard light footsteps nearing the bottom of the stairwell. They belonged to colleague Jiexiang Li, a fellow statistician. She was headed out and Mick stopped to hold the door open. Jiexiang also stopped. Which one should go first? They both smiled. She wasn't comfortable yet having doors opened for her. It was cultural. In China, she had explained once, whoever got to the door first opened it and went first. It was an equality thing. Mick chuckled as he entered the building. The two stopped briefly to exchange pleasantries at the bottom of the stairwell.

"You were not at lunch today," she offered. One of Mick's favorite pleasures was lunch with colleagues. This semester, teaching schedules caused the core of the group to consist of four full professors. The conversations were always relaxed and full of humor. None of them had

21

anything to prove. Although Jiexiang was new to the department, she often joined the older four for lunch, adding to the international flavor of the group. Dinesh Sarvate had grown up in India, Kathy Thom in Florida, Gary Harrison in Idaho, and Mick in Missouri. Well, slightly international, anyway. Jiexiang had taken some good natured teasing when it was noticed that there was a Chairman Mao ornament hanging on her key chain. She acknowledged the millions of deaths he was responsible for, but added that, even so, he was viewed in a positive way for making China a strong, world power.

"Today I had a craving for a good hamburger, so I went to Jack's," Mick responded to her observation. "What was today's hottest topic of conversation?"

"A show on Animal Planet. In the show, a bear cub was eaten by an adult male bear that was not the cub's father. I asked what other animals eat their own kind, and we talked about that."

"Jiexiang, you would have been the smallest one in the room. Fortunately for you, everybody must have remembered to bring their lunch." He smiled and feigned a sense of relief, and she smiled back.

She started to move toward the door and Mick started up the steps. The latter looked back as he climbed, slowing briefly to exchange hasta-la-vistas.

Mick was looking forward to getting into his office, getting a calculator out of his desk and doing some computations using the numbers he had written on a napkin. He had read the morning newspaper over lunch and was struck by the feature story on the front page, an article entitled *Citadel leads academies in sex assaults*.[iii] The Citadel and the College were both state institutions, both in Charleston. They also were arch athletic rivals in the Southern Conference. The former, with its corps of cadets, was one of the nation's service academies. In many respects, The Citadel was to South Carolina what VMI (the Virginia Military Academy) was to Virginia.

The two Charleston colleges were about a mile and a half apart as the crow flies, two if the crow drives. Because The Citadel (yes, *The* is part of the name, hence is capitalized) was a local institution, the headline made a tantalizing front-page lead. It seems that a survey had been taken at all of the service academies. "About one in five" female cadets at The Citadel responded that they had been sexually assaulted, whereas only 14% of the female cadets at all academies combined had so responded. Mick saw an attention-grabbing classroom example with a local flavor, one that would illustrate how it was easy to mislead people who were not trained to think statistically.

One of several general principles that Mick always hit hard when he taught elementary statistics was how blatant an offense it was to compare this year's number to last year's and then to reward or penalize somebody if there was a difference, without first having determined how much of a difference could be viewed as routine fluctuation. Without first knowing how much variation was endemic, no one had any business rewarding or penalizing people. People shouldn't be held responsible for numbers that were beyond their control.

The Citadel was a small college. He knew there were only about 1900 cadets altogether and that of these, female cadets made up a very small minority. In statistics, a small sample size always begat a large sampling error. In his pocket was the napkin on which he had written the relevant numbers from the newspaper article: 14%, 22 and 118. *About one in five* was a simplified rendering of the result that 22 of the 118 female cadets surveyed responded that they'd been sexually assaulted at least once.

Of course, this wasn't a perfect example of Mick's pet peeve – comparing this year's number to last year's – for he would use 14% as a population characteristic (what statisticians call a *parameter*), whereas he would view the 118 cadets at The Citadel as a sample. It was reasonable to treat 14% as a population parameter because it was based on a large number of female cadets having been surveyed nationwide. By viewing the 118 cadets as a typical sample of female Citadel cadets, he could then compute an interval estimate for the *true* population proportion of female cadets at The Citadel who would, over the course of many years' worth of samples, indicate that they had been sexually harassed.

Once at the office, instead of going to the lounge to get a cup of coffee, he unlocked his desk, got out a calculator and began keying in numbers. The "about one in five" was, more precisely, 22/118 = 18.6%. Hitting more keys and following a well-known formula, the professor found the customary 95% confidence interval for the true Citadel proportion: **18.6% ± 7.0%**. The swing of 7.0%, what statisticians call *the maximum error of estimate*, was large – close to what he had anticipated – because the sample size was small. Since 14% was between the interval endpoints of 11.6% and 25.6%, there was not a statistically meaningful (what statisticians call *significant*) difference between the sexual harassment rate at The Citadel and 14%. He decided this would make a great example to use when the subject of confidence intervals for proportions would arise in his freshman level statistics classes later in the semester. He would be sure to finish the example with a comment that the long-term sexual harassment rate at The Citadel might indeed be different than at the other service academies, but that no one could make such a call based on the data from just that one year. To make sure he wouldn't forget the example, he jotted the computations and some notes onto the napkin, opened the textbook to the section on confidence intervals for proportions, and put the napkin there. Deciding that it was too late in the day for coffee, he opened the package of M&Ms and began to let them melt in his mouth and not in his hand.

6 YOU STILL NEVER KNOW WHAT
YOU'RE GOING TO GET

His phone was ringing. It was Leomy, the math department's office manager.

"Could you help this person?" she queried.

"Aha. Someone in Charleston has a math question. Sure, put them on."

"Hi, Dr. Norton?"

"Yes, this is Mick Norton. How may I help you?"

"My name is George Hernandez. I have a math question. Can you help me?"

"I can sure give it a try. What's up?"

"The company I work for is doing something new. We're about to start soaking railroad ties in creosote. A railroad tie is eight and a half feet long, nine inches wide, and seven inches thick. We plan to soak a hundred of them at a time. The soaking will take place in a big vat, and we need to put the creosote in first. We don't want to put in more creosote than we have to. What we need to know is, measuring from ground level, how deep should the creosote be before we put the ties in?"

"OK, got it. What kind of tank is it?"

"Well, the tank is unusual. Imagine a big, really big, empty oil drum sitting on one of its ends. Then slice it in half vertically. Our vat is one of those halves. It'll sit on its curved side on braces so that the open top is horizontal, and shaped like a rectangle. Does that make sense?"

"Actually, you've given a really clear description of it. Yes, I understand. I need to make sure of something, though. When the drum gets sliced in half,

25

am I hearing you correctly that the vertical slice goes through the exact center of the circular top?"

"Yes."

"OK, got it. What are the inside dimensions of the rectangular top?"

"The inside diameter of the drum is ten feet. Call that the width of the rectangle. The inside length of the rectangle is 18 feet."

"Well, one tie has a volume of ... let's see ... that's 3.71875 cubic feet. So 100 of them would have a total volume of 371.875 cubic feet. I think I can see some inverse trig functions in our future. I would need to work some things out on paper and call you back. Also, there's an issue of these ties sitting in a container that has a curved surface. At first glance, I would think there should be just enough creosote to take up the air space that would have been between the ties if there were no creosote in there. But also, maybe we need to take into account how much creosote soaks into the ties. And maybe the biggest issue would be how 100 ties would pack together in a container like that. How do they go in? What kind of control do you have over positioning them in there?"

"I can find out and call you back."

"OK. In the meantime, I'll think about your problem a little and see if any other questions pop up."

Hernandez never called back.

7 APACHE FOG

He was listening to the radio as he drove home from work. The professor always looked forward to this part of the day. Talk radio was full of uppers – the latest antics of politicians. Just as the sun could be relied upon to rise in the East, some news items could be relied upon to show that many in the political class were really politically classless. Or was clueless apropos? Maybe both words together described it best. Of course, the math professor was confident that if he were to hold a political office, he would acquire no dirty laundry. On the other hand, maybe he, too, would succumb if exposed to that environment. Like wondering what comes after death, you can't know without being there.

He tried several talk radio stations and was disappointed. It was one of those days. Instead of reporting new outrages, they were recycling ones from the day before yesterday. Bummer. He went to Plan B, the golden oldies station. The song playing was not one he could remember ever having heard before. It didn't take long to be reminded that not all songs from the '60s would become golden oldies. So it was time for Plan C – listening to a CD. Part of the reason he looked forward to the drive home every day came from knowing that he had entertainment contingency plans.

The song the CD was playing sparked the recognition that he had an opportunity to earn praise from his wife. He and Libby enjoyed hits from the '60s. On occasion they would even brave to sing along. But Mick would rarely get the all of the words right. For him, melodies came naturally, whereas learning words was, well, hard work. Besides, he enjoyed the challenge of making up comparable lyrics to substitute for the words he couldn't remember. Libby, on the other hand, always seemed to remember all the words to everything. As he listened to the CD, he focused on learning the words. When the song was over, he started singing. "Celia, you're breakin' my heart. ... ."[iv] *Nailed it,* he thought when he was done. Well, maybe, of course. He surfed back to a radio station. The weatherman was describing current conditions and giving the forecast. A sentence caught Mick's ear.

"Be careful driving out there tomorrow morning because we will see Apache fog."

Apache fog. What a neat term. He'd never heard it before. Different kinds of fog could beset drivers. What kind of fog was he warning about? Apache fog might be a stealth fog, one that creeps in quickly and quietly, as if in a warrior's moccasins. It could go unnoticed until it was too late and the driver couldn't see squat. Yes, that could be it. Or would Apache fog have lots of clear spots with unexpected, dangerous patches of fog? Oh.

The thinking processes that bring about revelations are interesting. In this case, the subconscious was telling the conscious to wake up. His solving of the puzzle was accompanied by an embarrassed shake of the head and one of those barely audible, self-deprecating laughs that take place deep in the chest. The kind people do when they're alone. *How much time do we spend reacting to what we think we hear?* he mused. *Or to what we think we don't hear?* From now on, he would think *Apache fog* whenever he observed a Rosanna Rosanna Dana reaction – or had one himself.

He and Libby had dinner out. They did that a lot now. The kids were grown and out of the house, he worked and she worked, and neither of them had a homemaker inclination. They joked that eating out often was good for the economy. Their typical dining banter was trading workplace stories – and workplace gossip, naturally. And they ate well. So what was to criticize?

At home afterward, Libby went upstairs to her "office" – the room with her computer. She was a computer programmer who worked on projects for several companies. If deadlines meant that she had to work late at night several nights in a row, this was not a problem for her. Her stay-power and ability to focus were inspiring.

Mick sat down and began to grade the MATH 104 (elementary statistics) quizzes he'd given that day. It didn't take long. The students had done well, a good omen for the hour exam they would take on Friday. Or was it? Being prepared for a one-question quiz on a given topic didn't take much work. Being well prepared for a one-hour exam that would cover several chapters required a more willful preparation. And unfortunately, some freshmen could always be counted upon to assume that reviewing the night before the exam would be good enough.

The professor did one of his frequent self-audits, deciding that, yes, on numerous occasions, he had given his standard upcoming exam reminder. *Have you started reviewing for the exam yet? Remember that [fill in the topic of the day] will be on there, and that's tough stuff. Don't wait until the night before the exam.* Professors, like coaches, had to fire their charges up for a big game. He made a note to give the reminder again on Wednesday.

He double-checked the exam he had made up that afternoon. The questions were fair and clean. The number of questions was about right for a one-hour exam. *We'll know on Friday.*

8 MATH HUMOR, RIGHT HERE ON OUR STAGE

After reading the newspaper over a cup of coffee and a bacon-egg-and-cheese biscuit at McDonalds, he drove to school savoring the good feeling that comes from being caught up. Lesson plans were already done. Also, the department meeting had been cancelled. So in the afternoon, he could work on an invited paper he was to give at a statistics conference.

The morning would be devoted to reading new emails and to computer housecleaning. His email account at work, his desk at the office, and his desk at home all reflected the same packrat-like nature. This morning was one of those rare clean-up occasions. After reading the newly arrived emails, he planned to skim through all of the older emails and render summary judgments about which ones to keep and which to delete.

The only new email worthy of mention was a real day-brightener. It came from John Smith in response to a notice Mick had sent to former students informing them about a job opening that required applicants to know something about statistics and quality control.

> Hello Prof. Norton,
>
> It is good to hear from you. I don't know if you remember me but I was a MS Applied Math student taking your Statistical Process Control course (2001-2002) while stationed at the Navy's Nuclear Power School. After retiring from the Navy I took a position with GE in Atlanta as their Quality Assurance Manager / Six Sigma Black Belt for one of their Gas Turbine divisions. I have to thank you for the body of knowledge imparted because it

was a differentiator among Six Sigma Black Belts. In fact, I was asked to teach several of the SPC courses and still get calls from GE folks around the world requesting help with data analysis.

Currently I'm pursuing a Ph.D. with Walden University's School of Management in Decision Science and look forward to teaching someday.

Again thank you for your tutelage.

Regards,

Ben Doone-Goode

What an upper. One email like Ben's was enough to make his week. The satisfaction that came from getting emails like this one was what kept teachers teaching. Naturally, this one was a keeper.

He then began to skim the names of senders of old emails that had not been deleted. There were well over a hundred of them. Usually, just looking at the name of the sender was enough of a reminder for Mick to be able to make a quick decision on whether to cull. One email, though, took time because it caused him to reminisce. The email itself was nothing special, but the incident associated to it was a fond memory. In the late 1990s, Gordon Jones, then-dean of the School of Sciences and Mathematics at the College of Charleston, had asked Mick to sub for him at an event where minority undergraduates were to receive awards for excellence. Gordon was to present the mathematics and technology award. Instructions encouraged presenters to employ humor from their discipline when making award presentations.

Some students would contend that humor and math were mutually exclusive. *But that wasn't true,* he told himself. *Why just the other day, I created a definition that no one would ever see in a math book.*

When matrices A and B have the same dimensions and every entry in matrix A is greater than the corresponding entry in matrix B, matrix A is said to dominate B. In this relationship, A is called a domimatrix.

Mick mentioned the definition to fellow mathematician and colleague Kathy Thom, who promptly told him to go to his room.

The professor remembered being determined to try to incorporate math humor into the event. He had tried to remember any recent anecdotes he'd heard about math topics. Of course, he needed to choose one that anyone who didn't know much about mathematics could understand. And of course it had to have a G rating. And it hit him. He recalled an anecdote he'd heard not long before, one that he could adapt and make funnier. It was the only time in his life he'd told a joke to a theatre full of people.

He worked at length on the delivery and timing. Not since Jack Benny, Bob Hope and Walter Mitty had the world seen timing as great as this.

> "Event organizers told presenters that when we were up here, we could try to use humor related to our areas. I'd like to give it a go. Credit for the following story goes to Harold Reiter. I heard him tell it recently at a regional math meeting that was hosted here at the College of Charleston. Harold is the director of the North Carolina School of Sciences and Mathematics, a boarding school for bright students from across the state of North Carolina. One day he got a phone call from a woman whose son had been denied admission to the school." (PAUSE FOR EFFECT)

During the pause, he distinctly heard a female voice in the audience say "Oh, no." Several other female voices also came from the audience, expressing disapproval of the school's decision and support for the mother.

> "She wanted to know why." (PAUSE)

More voices came from the audience. "That's right," several of them said. The mother in the story had the backing of the audience. Although Mick couldn't see anyone out there because of the way the stage lights were directed, he thought it was a good bet that some heads were nodding up and down as they spoke in support of the mother.

> "After looking up records, Harold told the woman, 'Well ma'am, the long and short of it is that your son is in the bottom half of his high school class.'" (PAUSE)

An "Oh, no," and several other comments emanated from the audience. The professor sensed two tones in the voices. One was sadness. The other was disapproval of the explanation that the mother was given.

32

"The mother pondered this for a while before replying. 'Well, (PAUSE OF REFLECTION) I knew he wasn't in the top half of his class. (PAUSE) But still, (PAUSE) I had no idea he was in the bottom half of his class.'"

He recalled feeling relief at the audience roar of laughter – a sound spike really. When the laughter ended, he went on to say that in a world in which advances in science and technology came quickly, it was important for people to have a good understanding of mathematics and the sciences. So he was pleased to acknowledge a student who has excelled in mathematics, and so on and so forth.

Did I miss my calling? he thought. *That was fun. Was there a second career possibility – one of jazzing up other people's anecdotes? On the other hand, maybe going out on a high was a good way to end a brief stand-up career. After all, the event came with an audience that was ready to laugh.*

9 FACULTY LOUNGE TALK

The next day began with the usual school-day ritual. Mick would read the paper while eating cereal and fruit at home, or a sausage-egg-and-cheese biscuit at McDonalds, or a chicken or gravy biscuit at Chick-fil-A, or a waffle at Jack's. Libby was not big on eating breakfast, but sometimes would join him, particularly at Jack's. She liked the waffles.

This morning, it was waffles at Jack's. Jack's was a great place for breakfast or lunch (Jack closed up at 3:30 pm, sharp) and it was virtually across George Street from the campus. The place had the atmosphere of a 1950s diner, the food quality was good and consistent, and the prices were right. The consistency came from the fact that Jack was the only cook. Actually, if you sat at the counter, part of your dining experience was watching Jack, for whom cooking was an aerobic exercise. His movements at and between work areas behind the counter were constant, fluid, and dance-like. And mesmerizing. There was the French fry retrieval *snatch* done at the deep fat fryer, the hamburger spatula *flip* done at the griddle, the one-sweep mayonnaise *spread* at the work-counter, the batter-pitcher *pour* at the waffle iron, and many more motions. At the busiest time, lunchtime, his movements made for a kind of frenetic tai chi workout. Some customers referred to Jack's lunchtime movements as his *being in the zone*. The diner was best known for its hamburger platters. In fact, it was not unknown for Libby to order a cheeseburger platter for breakfast. This morning, they both ordered waffles. Mick also ordered one scrambled egg on the side, for the thrill of living life at the edge.

As usual, Libby left first, off to visit a client in Summerville, and Mick stayed to finish the newspaper over a topped-off cup of coffee. The next part of his morning ritual was to walk to the office and read the emails that had accumulated overnight. He soon came to Katrina's.

Dr. Norton,

Hey this is Katrina from your 11:00 am MWF class. I have a few questions and I am not going to be able to make it to your office hours tomorrow...that is unless you have another time I could meet with you. You said today that you would have a problem on there dealing with the range, variance, and deviation variance [she meant *standard deviation*] ...something else too....but I am having trouble matching up the words with the symbols. If there was anyway you could email me what they mean that would help a lot. Also..you said that there would be a problem on the test where you wanted us to enter it into our calculator and be able to figure it out from there. I am having troulbe remebering how to do it and I dont know if I am doing it correctly. I have a TI-82...and if that is also not a problem...maybe you could send me the directions to do that and what those symbols mean. Just to make sure I am studying right...do you just want us to go back through the homework and study all those problems? What else might help? If you cannot send me these things and you have another time where I could set up an appointment that would be great. I am just really busy tomorrow: I have Anatomy Lab at 8:00, a meeting with my Physiology teacher at 10:45 and then class at 2:00. Just write me back and let me know.
Thank you so much,
Katrina

Mick responded.

Hi Katrina,
You have let some things slide to the point where it'll be tough to recover. Give me a call so we can set up an appointment and assess your situation.

It was always tough to counsel the Katrinas of this world.[v] Her grades so far had not been very good, and her email revealed a history of not taking responsibility. The College had many fine students, and most knew what they were supposed to do. But there were always some who didn't seem to fathom their role in the learning process.

Getting a cup of coffee from the faculty lounge was the next leg of the morning ritual. The lounge was across the hall from the professor's office. Usually after reading emails, he'd go in and get a cup. He considered giving coffee a pass this morning. After all, he'd just had a cup at Jack's, so the craving was more for the ritual of having something warm to hold while making lesson plans than it was for need. From his office he heard lively discussion and laughter. That was the decision-maker. Off he went. The topic of the moment was a web site where students could evaluate their instructors. Anybody could look up any professor and see what the students there thought of him or her. Well, sort of. All faculty members knew that the tails of the distribution were over-represented on the website. More precisely, what was over-represented was the group of students who felt strongly enough, positively or negatively, to go to the website and write something.

There was regular coffee in a carafe on the burner, but only enough for a few sips. He made a fresh pot, noting that of the handful of faculty members who were enjoying the discussion, three were holding coffee cups. When the coffee level was low, it was becoming increasingly common for faculty members to pour themselves a *short* cup, being sure to leave a slight amount in the pot. They could then rationalize that there was no need for them, personally, to make a fresh pot (or to ask anyone how to make a fresh pot). When Mick was in a rush, he'd even done so himself. It helped his self-justification to rationalize that when he did so, at least he would turn the burner off. The lounge had experienced a rash of cracked carafes lately. The old saw about professors being aloof had truth to it.

"Did you ever look yourself up to see what the students say?" one asked another. Everyone seemed to have a strong opinion or a story. "How reliable can such evaluations be?" someone else asked. This gave Mick an opportunity to share his favorite outrageous story about reliability.

"Yeah, I've looked myself up, too, and here is my favorite evaluation. I don't remember the exact words, but the thinking reflected had to be like this:

> If you get him as a professor, run as fast as you can to the registrar's office and drop the course. He is the worst teacher I've ever had in my life in any subject. He cannot explain abstract ideas, is constantly over our heads, and doesn't answer student questions in class. Going to class is a waste of time. He is an embarrassment to the college. I don't know anything more about linear algebra now than I did before the semester started."

Most people were laughing or aghast at the lengthy string of criticisms. One professor's face registered real concern, though. Mick surmised that because the summation was so extremely negative, the English professor was wondering if there might be some validity to the negative critique. However, math department colleague Bill Golightly, the former department chair, actually had a puzzled look on his face.

"But you don't teach linear algebra," he said.

"Correct," Mick observed.

Actually, he had taught that course once, about twenty-five years prior to receiving the electronic *evaluation*, and again several years after telling the story. After he shared this story in the lounge, some colleagues who crossed paths with him on campus would stop to ask him how his evaluations were going in linear algebra, even people who'd heard the story second-hand.

10 ZANDRA

In the late 1970s, the Spoleto Festival came to the city of Charleston. Spoleto, Italy had hosted *the* Spoleto Festival for a number of years. This was an arts festival bringing together musicians, singers, dancers and other performers from all over the world. The performers put on operas, ballets, plays, recitals and so on, and people would travel great distances to buy tickets and attend the performances. The Charleston event was to be a two-week parallel activity at a second site. With the addition of Charleston, the Spoleto Festival was heralded as a *festival of two worlds*.

Ted Stern, then-president of the College of Charleston, had lent the College's support to the fledgling American version of the festival by hosting some of its events in college theatres and auditoriums, as well as outdoors on the cistern lawn. After several years, the event leadership could see the threshold that separated birthing pains from the arrival of a self-supporting, nationally known event.

One day, a baby grand piano was moved into room 200 of Maybank Hall. One door to the room was just outside room 218, Mick's office. On and off for the next several days, someone had been playing the piano.

Late May was a good time to do research. Typically at that time of the year, the professor's building would be quiet all day. The spring semester was over, and summer sessions didn't fill out classrooms like they would in later years, after the college would recognize the efficiency of having wall-to-wall summer sessions. Mick was spending a lot of time at the office trying to prove a conjecture, one that had to do with probability distributions. It was the kind of puzzle that, without a stretch, could be described as probability distribution forensics. Long before there would be a television series called NCIS, the professor was a kind of "Ducky," the character played by David McCallum.

Discovering mathematical proofs require focus, inspiration and, of necessity, quiet. Several days in a row, he had been deeply involved in mathematical thought when loud notes would suddenly emanate from room 200. Based on his several days of exposure, and on how long the guy had been fingering the ivories that day, Mick figured that if he were to leave for a bit, the guy might be finished when he returned. The professor decided to take a break.

Mick's office was a home away from home. Amenities were close at hand. The faculty lounge – a place to get a cup of coffee and have lunch – was just across the hall. As was the building's faculty men's restroom. The main math department office, with its administrative assistants, faculty mailboxes, and the department head was just around the corner and down the hall. If he was sleepy, Mick could close his office door, unfold a futon that he kept under the computer hutch, turn off the lights, and take a power nap.

He decided to swing by the math department's main office, check his mail, get a bag of M&Ms from the vending machine downstairs, then return and grab a cup of coffee from the faculty lounge. Sorting through the mail was easy. It was all chaff and no wheat. Getting the M&Ms meant more walking than anticipated because the vending machine downstairs was out. It was warm and sunny outside as he headed to the vending machines in Randolph Hall. Nearing the fountain, he crossed paths with Arlene Credenda. The two had met at a recent faculty reception. He couldn't remember what department she was in – one of the sciences, or maybe she taught stat for the sociology department or for the psychology department. But for sure, the two of them had hit it off easily and one thing he could remember was her name.

"Hi Arlene. I just finished reading an article my wife gave me. You might be interested in reading it, too. It's about differences in how men's and women's brains react to the same stimulus. They make such determinations after positioning electrodes at various places on the skull."

"Yes?"

"It seems that, for example, when posed with an analytical problem, different parts of the brain are energized. In men, the analytical region in the brain tends to react strongly, while in women, that region reacts, but not as strongly, while other areas react and supplement the analytical area. The indication is that men have an intense analytical focus when problem-solving, whereas women tend to have a more generalist approach, coming at a problem from different angles. The article describes how the brains of men and women tend to react differently to other kinds of stimuli as well. Anyway, you might find it interesting. I sure did. You're welcome to borrow it if you want."

"People shouldn't be doing research on topics like that."

"Why?"

"Looking into whether there are gender differences makes it more difficult to get across that men and women are equals. The issue is equity. Men and women should be treated alike. Research on topics like that interferes with a message that everyone needs to understand."

It's curious that she's not curious, mused the professor as he continued on his way. *I guess she won't be after me to borrow the article. I hope that kind of attitude doesn't come across in her teaching. We want students to be curious.*

After getting the M&Ms, he returned to Maybank Hall, got a cup of coffee from the faculty lounge, and stepped from the lounge out onto the porch, where he sat for a few moments. He was trying to think about his probability distribution problem. But the music, though fainter, was still noticeable. Even M&Ms hadn't kept his mood from going south. In addition to Arlene's having let ideology trump data, that pianist was still going strong. He had been patient for several days, but the musical interruptions were testing his limits. He knew that the college was providing space to the festival. Still, publishing was essential to his career. He had tried substituting other, lighter activities, figuring the piano guy was close to finishing. He wasn't.

Actually, the guy was playing much longer that day than usual. Mick had had enough. Nobody had checked with him to see if it was OK to move a baby grand piano into the room that was right by his office door. The College paid him to engage in mathematics, and this was interfering with his job. He rose and left his office to talk with the pianist.

When are you going to quit making that damn racket? was what he was thinking. Of course, he wouldn't put it that way to the guy face to face. *Part of my career involves doing research,* ... Mick was ready to begin.

As he left the lounge, there was Zandra. The always even-tempered Zandra, who was known as Alexandra Dengate only in official records, was the administrative assistant in the history department. She was standing just outside the door to room 200. At the sound of Mick approaching, she turned toward him. Her face had a look of peaceful rapture. "Isn't it beautiful?" she asked dreamily.

In response, he stopped and listened. And it was. Absolutely. It happened that the guy was a world-class pianist practicing for a recital. It had taken Zandra's prompt to get the professor to stop, take a deep breath, and listen. Listening – really listening – had changed what had been racket into soothing beauty. The beauty had been there all the time. What had been missing was a mindset willing to stop and look for it.

Occasionally in later years, he would smile and think about that day with Zandra, Arlene, the pianist, and mindsets. *How often do we misinterpret what we hear? Apache fog for sure*, he would muse.

11 MAYBE I'M JUST NOT GOOD AT YOUR TESTS

The professor expected that some of the day would be spent dealing with students who had not done well on Friday's MATH 104 exam. He had graded the exams over the weekend and returned them in class on Monday. Out of 38 students in his morning section, only three students had received an A. Eight had failed. Numbers were similar in his afternoon section. So Monday had provided another coaching opportunity. *Some of you clearly put in a lot of preparation for the exam*, he'd said. *But there were only three As. It would have been nice to have more. But some students didn't stay on top of the homework, and it showed. If your grade is not where you want it to be, what's a good strategy? For one thing, remember that right before an exam is not the time to learn material from scratch. Reviewing before the exam is the time to convince yourself that what you mastered earlier has stuck. The final exam is worth 150 points, and we'll probably have a couple of 10-point quizzes. So there are lots of points left to earn. Show me what you can do. PLEASE, stay on top of the homework. And if I wrote 'please come by to see me' on your exam, I meant it. Do come by. Oh yes, and did I mention to stay on top of the homework?*

Several students had called or emailed to set appointments. But Alice's email had Mick stewing a little.

> Dear Dr. Norton,
> I am a student in one of your Statistics/Math 104 classes. I was just wondering what type of test you will be administering to the class for the final exam, as in the format. Will it be like the tests you have already given us (questions you make up on your

own)? Or will it be questions taken from the book? I am making a C in your class, maybe even worse after the last test, and I find that I would be doing much better in your class if tests consisted of questions from the book, rather than questions coming from you. I think this is because in the book, they say "solve this binomial problem using..." and on your tests you simply ask for the answer. Even for the last test I studied for hours, making sure I knew every single piece of the study guide you gave us. Maybe I'm just not good at your tests, I don't know. :) I don't mean to criticize your tests, I was just wondering the format the exam will be in, so I know what to prepare for. Thank you for listening.
Sincerely,
Alice

Alice missed class too often, and Mick wondered if he was too abrupt in his response.

Hi Alice,
There is a difference between knowing how to work a question in a given section because of what was covered in that section and knowing enough to single out the right method when a test covers several sections. You need to rise to the higher level. Come to class and hear about the final.

As the day progressed, in addition to meeting with students, he continued working on a research question involving probability distributions. *Maybe today will be the day I crack it.*

At some point he went into the faculty lounge to get a cup of coffee. Fred, a professor of English, was alone and grading papers. A tall stack of bluebooks was on the table.

"Even bribery couldn't help the student who wrote this," joked Fred, pointing to the essay he was grading.

"How do you guys in the English Department do it? Reading all those essays and keeping your focus so you can do a good job of grading! I can't imagine doing it."

"I can't imagine grading math papers," Fred responded in kind.

"Well, I could have used your help last weekend grading some freshman stat exams."

"How'd they do? Did everybody get an A?" Fred wisecracked.

"The grades had what statisticians would call a uniform, or flat, distribution – about as many As as Bs as Cs as Ds as high Fs as middle Fs as low Fs. We all know, of course, some students just don't do well on tests."

12 CHARLESTON HOMICIDE DATA

The president of the College of Charleston had just instituted an event called convocation. In connection with the event, incoming freshmen had been sent a book. The idea was that reading the book would give freshmen at least one common background experience. Further, professors in topic-relevant courses were encouraged to capitalize on this experience by incorporating the book into their fall semester syllabi. The book chosen for the first convocation was *Everything You Thought You Knew About Politics...And Why You're Wrong*.[vi] The author, Kathleen Hall Jamieson, had been invited to campus to address students and faculty.

The committee that chose this book had asked some faculty members to read the book and propose thought provoking questions for the students of those instructors who planned to use the book. Since there were several chapters in which the author had interpreted the meaning of statistical data, Mick was asked to submit some questions.

One issue that happened to arouse his interest was a discussion of whether local television stations in Philadelphia were biased in how they had reported cross-racial crime. More black-on-white crime incidents had been reported than white-on-black crimes. This was presented as an indicator of bias. *Maybe they reported every cross-racial crime there was*, Mick thought, *and there was no reporting bias at all*. Why would anyone expect the two kinds of incidents to occur in equal numbers?

In addition to submitting classroom discussion questions to the committee, the curiosity bug bit him. He decided to look at data that he might reasonably be able to get access to – Charleston crime statistics – in order to determine if the numerical fallout would be similar to what had been reported on the TV news in Philadelphia. Blacks made up 43% of the Philadelphia population and 34% of the Charleston population, not a perfect match. Still the data might be illuminating. The City of Charleston Police Department was kind enough to provide the professor with a table of homicide counts covering the most recent 6 and 1/2 year period (see Table 1 for the breakdown of homicides by race and sex of perpetrator and race and sex of victim – B, W, O, M and F denote black, white, other, male and female). Also included in the table were unsolved homicides by race and sex of victim.

The Charleston newspaper frequently ran articles on crime issues, including the once-a-year report on whether homicides were rising or falling. Typically, the homicide article would report the percentage increase or decrease in the number of homicides from the previous year. And every year, the professor would frown when he read the article. Drawing inferences by comparing this year's number to last year's, without having any idea about how much of a swing fell within routine fluctuation, was a classic mistake made by those untrained in statistics. If the amount of routine variation – typically measured by the standard deviation – were an unknown or could not be estimated, it would be impossible to know if a given one-year swing was significant.

The paper also had run occasional articles on black-on-black crime being a serious problem in Charleston. But to best of Mick's memory, numbers that quantified the extent of the problem had not been provided. He decided that the Table 1 provided an eye-grabber example, one in which data from the college's own city could be used to help students understand conditional probability. Using real data to help students understand how the world worked was one of the paradigms of teaching statistics. The textbook's examples and homework problems provided plenty of similar tables around which he could have built the lesson, but there was nothing inherently memorable or captivating about them. And so, he added the Charleston homicide data to several other real-life data sets and examples that he used to help illustrate important concepts in introductory statistics classes. The other examples had come from personal experiences in expert witnessing and from consulting projects with manufacturing companies. One thing he particularly liked about the new acquisition was that the homicide data provided insights about Charleston itself.

Who Kills Who in Charleston by Race and Sex
January 1, 1996 – June 30, 2002

Perpetrator   Victim

	B/M	B/F	W/M	W/F	O/M	O/F
B/M	30	4	2	1		
B/F	1					
W/M	1			3		
W/F						
O/M						
O/F						
Unknown	10	1			1	

Table 1 – *Source* : City of Charleston Police Department, 2002

In learning how to use a table like this one to compute conditional probabilities, his students would also learn several things about Charleston that were true for the 6 and 1/2 year period.

The first was that the victim was black in 11 of the 12 unsolved homicides (Perpetrator Unknown). The first time Mick had presented these data in a class, a student suggested that police probably didn't try hard to solve such homicides. When he used the material after that semester, Mick would relate what that student had opined, and follow by asking the class to identify another conditional probability that would either help support or help contradict the student's contention. Students would almost always decide to compare 11/12 to the probability that the victim was black when the case *was* solved. Since that answer, 36/42, was similar to 11/12 (36/42 is between 10/12 and 11/12), students would see not only that there was no evidence to support the police-don't-care theory, but also that the really big picture was that roughly 11 out of 12 homicide victims were black, period.

Of the four homicides committed by white males, three victims were white females, whereas females were victims in only one-sixth (9 out of 54) of homicides generally.

Of the 42 solved homicides, 38 had black perpetrators, and of these, 30 involved black males killing other black males. So the data during the period indicated a contrast – white guys tended to kill white females whereas black guys tended to kill other black guys.

The data also made it clear, he would point out, that women apparently didn't kill people like men did (only one of 42 known perpetrators was a female). Mick would point out to students that his wife disagreed. She maintained, with a smile, that women were smarter, that they were responsible for all of the unsolved homicides, and even for the unknown homicides, the ones with victims too well hidden to be discovered.

13 SHAKEN, NOT STIRRED

The first thing he would do today would be to choose two volunteers from his class of teachers and lead them through a demonstration of taking a *random walk on the number line*. The exercise (the best puns are unintentional) was one of several he used to get across the concept of average behavior over the long-term when a game is played repeatedly. Not only would the activity help the teachers get the concept, the teachers could use the activity with their own students – mostly middle schoolers – to help them understand.

Knowing what to expect as a result of long-term aggregation was another one of those *help people understand how the world works* concepts in statistics. Although his classroom activities often involved cards or dice, he also would show his charges real-world applications, all of which were gambling, even though cards or dice were not involved. To illustrate, suppose someone pays an insurance company a premium of $300 for a one-year $10,000 term-life insurance policy. Insurance company actuaries study the statistics and probabilities of the industry. If they know someone's age, sex, whether they smoke, and so on, they can look up the probability that a person with those characteristics will survive the year. They use the information to set premiums. Naturally, some policy holders will die during the year and cost the company big bucks. But the vast majority of players, oops policy holders, will survive the year, and their $300 per head can be viewed as pure profit for the company. Insurance companies do not exist to be revenue neutral. The premium charged was arrived at with one purpose. When the gains and losses are averaged across *all* policy holders in a given year, the company will, barring a catastrophe, make a profit. In fact, the students taking the course

49

would learn a straightforward method of how to project what the company's average profit per policyholder would be. Also, given a number X, they would learn how to answer a question such as *What is the probability that the projected average profit and the actual average profit will turn out to differ by X dollars or more?*

For the reader's information, the professor could be counted upon to tell his students that well meaning, politically correct do-gooders who tried to force companies to charge unisex premiums drove statisticians nuts. Ignoring known, statistically relevant factors when setting premiums would necessarily result in one demographic group helping to subsidize another. There is no doubt that people would materialize who, misguidedly wrapping themselves in the banner of fairness, would argue that a gerbil and a box turtle should pay the same amount for a one-year $10,000 life insurance policy. After all, since either one of them could die in its first year of coverage, why should a gerbil pay a higher premium? *You could always find examples of just about anything,* Mick would respond, *and you shouldn't let extreme cases be the defining factor when formulating general policy. You might have special rules for how to deal with special cases, but general policy should be based on global truths.* The typical box turtle lives more than twenty times longer than the typical gerbil. Everybody knows that's why gerbils are charged higher premiums.

Ah well, back to random walks. The classroom floor was tiled with one-foot by one-foot vinyl squares, a perfect layout for an *x-y* axis system or for a number line. Mick stuck a post-it note in the center of one of the squares in the walkway that cut the room down the middle from back to front. After announcing that that square was the origin, he called for a volunteer who would perform the random walk. Tammy, one of the more socially outgoing students, eagerly responded. The professor directed her to stand on the origin. Lauren, almost as live a wire as Tammy, volunteered to the call for someone who would roll the die that would determine Tammy's movements.

He then laid out the rules for Tammy's random walk. If Lauren rolled a 1 or 2, Tammy would move forward (toward the blackboard) one square; if a 3 or 4, Tammy would move forward two squares; if a 5 or 6, she would move backward three squares. Taking advantage of the probability topics they had been working on for several days, the professor asked the students to make a best guess as to where Tammy should be expected to be after fifteen rolls of the die.

Actually, it was possible to arrive at the answer without knowing anything about expected value – the relevant topic. One just needed to apply a little common sense. In fact, the professor used the common sense approach to shed light on why one of the formulas that the students learned was easy to remember. While it was true that Tammy's movements would be random, it was also true that in 15 turns, there ought to be five turns, give or take, on which she would move forward one square, and five turns, give or take, on which she would move forward two squares (that advances her 15 squares toward the board altogether, give or take, right?), and five turns, give or take, on which she would move three squares away from the board. So, give or take a few squares, one should expect to see her still near the origin, with the origin being the best single-position prediction of her final position. Seeing how their predictions would square with the actual outcome was part of the purpose of the activity.

Because things could get out of hand when middle school boys rolled dice on desktops, Mick had a preferred method for showing how teachers should have their own students toss dice in a classroom setting. In fact, he had had the teachers use it two days earlier when he involved them in several simulation activities. Dice were issued to students in cardboard coffee cups. A student would take the cup in one hand, cover the top with the other hand, shake the cup vigorously up and down at least four times, and then peek into the cup to see the results of the toss. One of the teachers had even commented that she would abandon the dice tossing method she had been using with her students. This, of course, was an upper for Mick.

The professor issued Lauren a cup and a die. The time had arrived for Tammy to execute the walk as Lauren specified the marching orders. Everyone had their eyes on Tammy as Lauren began announcing the results of the die tosses. "Two," Lauren called out, and Tammy moved one square toward the front of the room. "Six," she called out, and Tammy moved three squares backward. "Six again," she announced, and Tammy moved three more squares toward the door.

Two sixes in a row, thought the professor. *Well, it happens.* He was a little worried that with only fifteen tosses, Tammy's final position might not reassure the students about the reliability of their predictions. This was always a risk in hands-on activities in which class-time limitations required putting limits on how many repeated trials could be performed. *Let's hope for a slowdown on sixes.*

"Five," Lauren announced, and Tammy moved three more squares toward the door. "And six again," she observed in a tone of amazement. *Would Tammy move out the door and into the hallway as the random walk progressed?*

51

Sometimes, classroom simulations that involved probability could go slightly awry. But this was nuts.

Having images of being near the roulette wheel in Rick's Café and watching Humphrey Bogart tell the Bulgarian to keep betting on 22, the professor turned his attention from Tammy to Lauren. As if she were handling a mixed drink, Lauren was swirling the cup in gentle, horizontal circular motions, which would have had the die sliding around against the inside wall of the cup. After five swirls, she tilted the cup slowly until the die slid gently out onto her desk just at the edge of the cup brim. "Six," she announced as she placed the die back in the cup with the same orientation in which it had slid out. She started swirling again. *Well, she got the part right about kids causing disruptions by rolling dice off their desks.*

"Lauren, do you remember how we talked about independent events the other day? When tossing dice, we observed that in practical terms, independence means that the outcome of one die toss does not influence and isn't influenced by the outcome on any other toss. Remember? Recall how on Monday everybody gave their cups at least four good shakes up and down?" Interestingly, Tammy's ultimate position was the spot she was at when Mick had Lauren change her technique.

And so, the concept of independence was revisited and reinforced. In fact, the students probably would remember the activity and concepts better than they would have otherwise – a silver lining that far outweighed the cloud.

14 SOLDIERING PROBABILITIES

It goes without saying that random events can have a life altering impact on a soldier. The professor would think of this every time he and one particular friend would get together. Jim Seibert and Mick hadn't traveled in the same circles when they were in high school. Jim was a partier who wasn't much interested in schoolwork. But they'd seen each other socially at a couple of class reunions over the years, and it was easy for them to connect because they shared two strong bonds. One was wrestling. They'd spent several years in the same weight class and had wrestled each other many times, creating a host of shared memorable moments. Two athletes who respected each other and who went all out to beat each other, especially in a sport as grueling and intimate as wrestling, shared a special bond. Vietnam provided the second bond. They had not been in-country at the same time, nor even done the same things. But there was a bond nonetheless. Former soldiers always have a warm appreciation and respect for each other – a special, instinctive bond. Even if they are from different generations, they are brothers.

Libby and Mick had attended the NCAA wrestling championships many times. One year when they happened to be held back home in St. Louis, a work project that arose on short notice kept Libby from going. Mick called Jim and asked if he wanted to go to the semi-finals and finals. He did, and after the finals they went out for a beer.

It would become a habit that when Mick was in St. Louis, he and Jim would have dinner and a beer, reminisce about high school wrestling, and trade Vietnam stories. In all things athletic, Jim had an aggressive, winner's instinct. Perfectly befitting his highly competitive attitude in athletics and personality, Jim became an Army ranger. He earned medals in Vietnam – two purple hearts, three bronze stars and one Silver Star. The major combat he saw had affected him negatively ever since. Waking up in the middle of the night in a sweat after dreaming about snakes, for example, was one of a number of post-traumatic stress issues. Talk about random events affecting a life.

Admittedly, Mick took some amount of pleasure in having received three Army commendation medals during his tour. Mostly, though, he was pleased with the third one, which, as it happened, was a consolation prize. MSgt. Matheny, an NCO he had worked under, had nominated him for a bronze star for service. But the higher-ups who called the balls and strikes decided to instead confer ARCOM number three. Counting awards was not something he thought much about. Mick figured he was just doing his job. Actually, he would have been hard pressed to describe to anyone exactly what it was he had done to earn the medals. Regarding the first two, all he knew was that he must have made somebody in the First Infantry Division happy. What did matter, though, was Matheny's opinion. It was that third ARCOM in which he took the most pride.

The critical thing here, though, is that Mick's ARCOMs were all for service. When it came to trading stories with Jim, Mick was most definitely trading up. A typical Jim-story would be brief.

Jim was a squad leader and his unit had been on the trail of an NVA unit for several days. His squad had lost men. One day, the quarry was atop a hill that had been defoliated. A lieutenant had just joined the unit and told Jim to take his squad up the hill and finish the enemy off. Jim pointed out that an air strike or artillery would do the trick instead, and that on the defoliated hill, he would expect to lose men along the way. Some discussion ensued. According to Jim, the lieutenant assured him that he had studied tactics, was well trained, and knew what he was doing in sending the squad up the hill. After some additional back-and-forth, the lieutenant told Jim to follow orders. More discussion followed, of the heated variety.

Typically, new NCOs and lieutenants who had never been in combat and who were assigned to take charge of combat units would, until they got their feet wet, seek the advice and counsel of those who, although below them in the chain of command, had combat experience. Not this guy. Jim refused to follow the order. The lieutenant brought charges under the UCMJ (Uniform Code of Military Justice). Jim had failed to follow a direct order. He was guilty of course, but his CO told him not to worry about it, that another lieutenant had witnessed the discussion, had reported what he'd observed, and that the new lieutenant was being, ah, transferred.

Several years after telling Mick this story, Jim would concede, sheepishly, that there actually was a little more to the story. With the squad leader absolutely refusing to put lives at risk needlessly and the lieutenant badgering him with an order to take a squad up the hill, the discussion escalated to the point that the officer, acting with the full fury of higher ranking authority, got in Jim's face. A mistake. Jim decked him. Jim's admission conveyed not the slightest hint of regret. Actually, he described proudly how the blow – or was it blows – had been very well delivered. There was to have been a court martial, but the charges were downgraded to an Article 15 under the UCMJ – similar to a felony being downgraded to a misdemeanor.

Jim had lots of good stories, all fascinating, all brief, and all revealing him to be close to a real-life version of the brawling, lovable characters played by Victor McLaughlin in those John Wayne movies from the 1940s and 1950s. There was the time Jim was in a bar while on R&R in Australia, getting along well with an Aussie over a beer. His undoing was to think he was paying the guy a compliment by telling him he was a "good Limey." The fight that ensued brought the authorities, who hauled them both off. Jim concluded the story with pride by pointing out that the two of them became friends while in custody.

And there was the time at Fort Hood, Texas when he and some other seasoned combat vets just back from Vietnam didn't return to base in a timely fashion after being gone on a pass. Part of their decision-making involved the belief that making morning formation was stateside silliness – not all that important. He got busted two grades from sergeant to PFC (E-5 to E-3). Soon thereafter, however, he was promoted to E-4 so that he could be awarded his silver star without undergoing the indignity of having everyone present know he'd just been busted two grades. The Army had been one big, random event that affected Jim's life.

Mick would share stories back. The following example was long winded, but Jim enjoyed it, mostly because it was about the soldier overcoming the system.

It was near the end of Army basic training at Fort Leonard Wood, Missouri in the spring of 1969. When it had been determined what kind of specialty training each basic trainee was to receive, but before the trainees themselves knew, those who were to go to infantry AIT (advanced individual training) were separated out for familiarization with the M-16, the weapon used by infantrymen in Vietnam. Up to that point, their rifle training had been with the M-14. None of them understood why they had been singled out to learn about M-16s. The physical aspect of basic may have made them too tired to think it through.

After becoming adept at taking an M-16 apart and putting it back together, they were shown how to zero one in. Each was then assigned a weapon and zeroed it in. The M-16 captivated Mick. With its plastic stock, it was markedly lighter than the M-14. It had a carrying handle and a pistol grip. After his having lived with an M-14 for weeks, the M-16 had the appeal of a toy to a kid. It even looked a little like a kid's toy. Another reason for the fascination was that it was the weapon that American soldiers had been using in Vietnam, and he thought that he might possibly soon be one of those guys. Honestly though, the stronger of the two feelings was that of a kid with a new toy. In fact, the trainees were even told that the design was based on a toy rifle that some general had seen, and that some of the plastic parts were made by Mattel. The Mattel thing sounded plausible at the time, anyway. Gee, could some of the things trainees were told have been exaggerations?

After zeroing in the weapons, trainees were divided into pairs and sent to the firing range. Each scored the other on hitting pop-up silhouette targets ranging anywhere from 50 to 400 meters away. They were to try to hit every target and replace empty magazines quickly enough to keep pace with the targets. Enough targets would pop up to consume three magazines. A number of trainees had come to the same conclusion: *I'm not sure*

I want to look good with this weapon. Why, they might send me to infantry AIT and I'll go to Vietnam.

In later years, Mick would feel embarrassed for having considered that line of thought. But then, as a draftee who was only seven weeks from having been a civilian – and for sure, being in the Army wasn't his idea – he wasn't exactly ready yet to embrace a duty-honor-country mentality. In later years, he would smile and take solace in the biblical observation that kids think and act like kids – always have, always will. Anyway, back to M-16s.

It was at this point that Mick learned something about himself. He could have aimed to miss – a sure-fire strategy, so to speak. Instead, he took a probabilistic approach. He decided to aim at the very left edge of each pop-up silhouette, as in target in the right half of the sight and landscape in the left. After knocking down the first three targets, he concluded that the weapon had to be shooting a little to the right. So he started aiming at the right edge of each target. The next three silhouettes went down.

Several things became clear at that point. He must have done a good job of zeroing in the M-16, it felt very comfortable in his hands, and he was doing a lot better with it than with his M-14. Also, he was having fun. *Let's see what I can really do with this*, he thought.

Taking distance into account, he began aiming at a point on the imaginary vertical line that passed through the center of each target, thereby illustrating that a kid can be counted upon to forget the prime objective when he's having fun. He missed two targets, one of them 400 meters away, in three clips.

"I can see why they picked you for this," said the partner who scored him. Mick often wondered if his partner was not a very good marksman, or if he was just a smarter guy.

However, he felt a calm of relief in having done the right thing after firing the first six rounds. And it wasn't because stat folks want good data. Also, that he had begun with the left edge-right edge strategy[vii] told him that when he chose to not excel at something, it was not in his nature to be blatant about it. After all, not excelling at something was not as embarrassing as looking absolutely hopeless at it. A corollary was that if he really wanted to look like he didn't excel at something, he needed a more willful strategy to overcome chance.

Fast-forward one year to an office job he had during his last few months in Vietnam, where personnel management, not infantry, was his day job, despite having gone to infantry AIT. He worked with MSgt. Matheny and SSgt. Cox, both career-Army, and both good men. There were just the three of them in the office. A Spec 5 (E-5) by this time, Mick was the lowest ranking of the three. The officers in immediate charge consisted of a captain and a warrant officer. Rumor had it that the former was made an instant captain because he had a law degree. He wore his captain's bars heavily, as the expression goes. Basically, he had difficulty relating to enlisted men. His in-charge demeanor was overdone, and he was condescending, pompous and smug in giving orders.

One morning, SSgt. Cox and Mick were in the office when the captain entered. In explaining to Cox a task that needed to be performed, the good captain exhibited all of the descriptors just mentioned. He left, after giving an admonishment that it had better get done. The instructions had the over-explanatory, stern tone that a mother would use to get a rebellious four-year-old to do something.

After the captain left, the two of them just stood there in stunned silence.

"Why does he talk that way?" Mick asked. "Does he think people won't follow his orders?"

"I don't know," said a seething Cox. "But one day, he and I are going to have words."

One morning several days later, the captain announced that he needed a driver, someone who could chauffer him to Saigon when needed. A military driver's license was required in order to drive a jeep in Vietnam. The captain gave Mick two mornings off from work to take a course that would earn him the license.

Later that day, Mick mentioned to SSgt. Cox that the idea of being someone's driver, particularly this captain's driver, had not exactly made his day.

"You could always fail the test," quipped Cox. "And there isn't anything he could do about it."

It didn't necessarily sound like Cox was making a suggestion. Maybe it was just a wisecrack. But it did start the mental wheels turning. Up until being in the Army, doing well on tests was how the future professor had measured success. At that un-advanced stage of his life, failing a test intentionally was not an idea that would have crossed his mind. SSgt. Cox helped him grow.

The first morning of the course was spent in a classroom. An instructor explained the nomenclature of a jeep, what elements were in the drive train of a jeep, how many quarts of oil a jeep held, and so on. Similar information was presented about a deuce-and-a-half truck. At the end of the morning, they were given manuals. That night, each soldier taking the course was to review the notes he had taken and study the manual. A written test would be given the next morning. After passing the written test, they would get driving experience and finish with a driving test.

Mick decided to fail the written test. But the experience of using an M-16 on the firing range had taught him to not leave anything to chance. After returning to his hootch (quarters), he put the notes and the manual in his locker, locked it, and vowed not to open the locker that night. In studying, he would have risked answering too many questions correctly. He socialized that night at another hootch.

The written test the next morning was an odd experience for Mick. At that time, he had a bachelor's degree in mathematics, another in math education, and a minor in physics. He had graduated magna cum laude; been drafted out of graduate school in mathematics at the end of his first semester at Oklahoma State University, where he had been awarded an NSF traineeship with no duties. And he had received very high scores on the battery of tests the Army gave trainees during the first week of basic training.

None of this helped him remember what elements were in the drive train of a jeep. All he wanted to do was fail the test; he didn't want to look like a complete moron. Which was about to happen. He racked his brain to remember even the most basic of pure-memorization facts about jeeps, ones that the instructor had presented so clearly the day before and which he'd made absolutely no effort to memorize. He couldn't remember. How many quarts of oil did a jeep hold? Who knew? He guessed at just about every question. The clock was ticking and he was in a panic mode.

Ultimately of course, time did expire and the students queued up to hand their exams to the NCO in charge, who graded each exam on the spot, in the presence of the test taker. That same NCO had made some excellent presentations the day before. It was clear that he had been at one with cars and trucks his whole life and that he really knew what he was doing. He stared at Mick's answer sheet for what seemed like hours, in silence. Mick had never felt so embarrassed. "Son," said the NCO, "you missed the easiest questions on the test." The tone conveyed that he wondered how Mick had managed to successfully find the way to the motor pool. "You can't do any more today. You'll have to come back next week and try again."

On Thursday of that week, the good captain mentioned that he needed to go to Saigon on Friday and said that Mick should make arrangements to get them a jeep.

"I didn't pass the test, sir."

"Oh, that's too bad. Don't be discouraged. It's not unusual to mess up the driving test. It can happen to anybody. You can take it again next week. Did you back into something? What did you do to fail?"

Now there's a question I can answer in different ways, thought the Spec 5. "I failed the written test, sir."

"The written test! How could you fail the written test? Well, go back next week and do it again."

Mick moped as he called to sign up for the next week's course. *It's not over*, he thought. Further, that NCO who taught the course would see him again. But in the couple of days since he'd failed the test, the powers that be had decided there must have been a need to strengthen the course, which they did by requiring a five-morning course. The good news was that, in light of how much time Mick had left on his tour of duty, the office couldn't afford to let him be away every morning for a week. So it really was over.

A few days later, MSgt. Matheny took Mick aside and told him that the captain suspected that he'd failed intentionally. Matheny's comment was a mix – half FYI, half I'd really love to know.

"Sgt. Matheny," he responded with apparent horror, dismay, and just maybe a twinkle in his eye, "what a terrible accusation to make." And that was that.

Jim enjoyed Mick's stories – the beer may have helped – and Mick was awed by Jim's. Mick was proud to know Jim and to be in a position where he could trade up.

15 MACABRE JOB PROBABILITIES

After getting drafted out of graduate school in mathematics, Mick had done Army basic training at Fort Leonard Wood, Missouri, then infantry AIT at Fort Lewis, Washington. After infantry AIT, he was sent to Vietnam, where he was assigned to the First Infantry Division. It was there that someone noticed that he had graduated from college and had high Army test scores. Maybe most importantly, someone had entered in his personnel file that he had a secondary military occupational specialty – one in personnel management. Where on earth had that come from? Mick hadn't the foggiest. In any case, instead of life rolling the dice in his behalf and issuing the most likely outcome, his being an infantryman out in the boonies, he was assigned a support job – one that turned out to be very weird – in Xi An (pronounced ZEE ahn), the location of Headquarters Company of the Big Red One (the First Infantry Division). Actually, it would be inaccurate to say that winding up in Headquarters Company was all luck. He had input. During an interview, a warrant officer had told him that Headquarters Company needed someone to maintain some rosters. He gave Mick the choice of being assigned to the company and learning the job as he performed it, or being assigned to a combat unit and do what he had been trained for.

Life experience prior to military service doesn't give people a clue about some of the jobs that exist in the armed forces. When it was explained to him that he would be maintaining some rosters, he expected that the job would involve routine office paper shuffling. It wasn't until he started performing the job that he realized its bizarre nature. Mick worked with a small group that maintained PIRs (personnel information rosters). Every military unit had a structure, and each soldier in the unit occupied one line in that unit's structure. Depending on the kind of unit, there were lines for cooks, platoon sergeants, riflemen, mortarmen, and so on. The PIRs he kept were for infantry units and an attached artillery unit. Each infantry battalion had so many companies, each company had so many platoons, each platoon had so many squads, and each squad had so many riflemen and so many grenadiers in it. Mick would read the morning reports of the units whose PIRs he kept, see who was a KIA (killed in action) and read the brief incident description (e.g., platoon was seeking an unknown size enemy force), see who was an MIA (missing in action), who was vacating a line because his one-year tour was up, and who was being transferred.

It was the time before people had computers at their desks. He would take out a ruler and pencil, find the right PIR, find the right page, find the line with the soldier's name, line him out, and note an explanation in the margin – e.g., KIA. Mick preferred a red pencil to line out KIAs so they'd be easier to find later. There also were occasions, not all that rare, when a morning report would say that someone had been killed in an auto accident back in the states after having gone home on emergency leave. The typical reason for a leave was to visit an ailing parent or attend a funeral. While enjoying their brief freedom from danger, did they drink too much and then go driving? Morning reports didn't explain those details.

Also coming across the desks of Mick and his work group were personnel folders of newly arrived enlisted men who were being processed into the First Infantry Division. The new arrivals were to be assigned to unoccupied lines in the PIRs. *Let's see now, your MOS (military occupational specialty) is 11-B-10 (infantryman). What squad of what company was it that lost three riflemen in an ambush a couple of days ago? Oh, yeah. There it is and there you go.* Mick would pencil a name onto a line, and orders would then be cut (prepared) that would send the soldier to the appropriate unit. *What a weird day job*, he often thought. The first few days on the job seemed surreal, in part perhaps because his primary MOS upon arriving at the First Infantry Division was 11-B-10.

It was possible to acquire a morbid sense of humor there. When Mick was in high school, he'd once asked the uncle of a friend of his what he did for a living. "I'm a short order cook at a crematorium," the uncle had responded with a chuckle. Mick couldn't understand how anyone would see humor in a wisecrack about death – until he'd started wielding his red pencil. He remembered having lined out four guys from the same platoon in one day once. *I guess they found that 'unknown size enemy force' they were looking for,* he remembers wisecracking to himself. Decades later, seemingly all too often, he would be stopped at an intersection, glance in his rear view mirror and notice that a KIA was right behind him. *What a lousy name for a make of car. That I always seem to notice when a KIA is following me must be some kind of penance for those wisecracks,* he would wisecrack to himself.

Each soldier in the office did guard duty when it was his turn, which was nearly once a week. This meant spending many a night on guard duty on the perimeter, either at a bunker or in a tower. The towers were just like what you'd see in the movies. Usually, and always whenever Mick had guard duty, nothing happened. He also participated in a perimeter sweep once. A group went outside the wire just before dusk, weapons locked and loaded, looking for any bad guys who might be there. They swept in a line perpendicular to the perimeter, men about 15 feet apart. As the line advanced slowly around the perimeter, everybody's eyes were big. He could never remember having been so focused – being on the lookout for bad guys. And for snakes. They stopped once and everyone knelt in the tall grass. Well, everyone except the guy to his immediate left, who stood there like a sightseer. He made a big silhouette. Mick caught the guy's attention with a slow, downward arm motion and got him to kneel. Naturally the sweep produced no bad guys to shoot. *Cause and effect – I'm here,* Mick noted.

One night, the future professor was dispatched on short notice to an intermediate bunker. Intermediate bunkers usually went unmanned. But on this night, there was a firefight in the distance and every bunker was manned in case action moved to near the perimeter. Red tracers (ours, as Mick would say) were going from one thicket of trees into another, and green tracers (theirs), were coming back the other way. It had a captivating beauty, this strange fireworks display.

Although not in the firefight, they were close enough that the bad guys could try for a lucky shot. One soldier Mick worked with had a bullet whistle by his ear. Once it became obvious to the guy that he was at the margin of being within range, he decided that watching the fireworks display from a standing position, even at night, wasn't such a great idea.

As Mick jogged with his M-16 and magazine-laden bandolier on the way to the bunker, there was an explosion to his left, maybe 50 feet away. He froze momentarily and growled his favorite four-letter word. It could have been an RPG coming from that distant thicket. He did happen to be the only guy jogging down the road at the time. But his mind was on getting to the bunker, and he hurried on, not thinking any more about the explosion. It wasn't until years later that he would think about it and wonder. Maybe he had provided a distant silhouette. Or maybe it was just a random shot to inside the perimeter. As to what it was that exploded, who knew? Maybe someone had been aiming at him – personally. Maybe not. He would never know. But he would wonder.

Certainly, he'd never had occasion to shoot at anybody. Insofar as his war stories went, that's all there was. His day job was at a desk. The Headquarters Company at Xi An even had a swimming pool. After Nixon Vietnamized the war and pulled the First Infantry Division out of Vietnam, he was assigned to the 527th Personnel Services Company at Bien Hoa (where he worked with MSgt. Matheny and SSgt. Cox). Life was tougher there. There was no swimming pool.

Not yet mentioned as an example of life rolling the dice in his behalf is that, from infantry training and periodic trips to the firing range to practice, Mick developed tinnitus – high pitch hearing loss – for which, many years later, the VA had given him hearing aids. John Thom, Kathy's husband, had once mentioned to Mick that it must have been demoralizing to notice his hearing deteriorating over the years. The response was relaxed and matter-of-fact.

"Fifty-eight thousand guys got dead and I didn't. When you look at it like that, some hearing loss isn't that big a deal."

Every now and again, the professor would think about those 58,000 and the fact that he wasn't in the group. He hadn't been out in the field doing what he was trained for, but instead contributed to the effort with a support job. So it wasn't guilt, exactly, that he felt. All he knew was that, sometimes, he'd think about them. And about probabilities.

16 AN ARGUMENT LARRY SUMMERS COULD HAVE USED

After reading emails in the morning, Mick checked out his usual web sites. One of them contained an article by Linda Chavez entitled *The Shibboleths of Academe*. The piece was about the president of Harvard, Larry Summers, who had given an address in which he'd observed that there was under-representation of women in the sciences. Harvard faculty were on his case for having had the audacity to wonder during his address whether the inclination to go into the sciences might be due to inherent differences between men and women. Ms. Chavez said that Summers ought to have been able to ask such a question without being ganged up on. The professor sent her an email.

Subject: Shibboleths

Dear Ms. Chavez,

I just read your piece about potentially innate differences between men and women and a feminist scientist incensed at what Harvard president Lawrence H. Summers said (*The shibboleths of academe*). It reminded me of two conversations I had with woman, fellow mathematician and good friend, Ewa Wojcicka. Ewa (pronounced Ava) died as the result of a tragic auto accident about 10 years ago. She came from Poland, where she had been treated from childhood as a highly talented athlete. In the U. S. she focused on academics and earned her Ph.D. in a branch of mathematics called combinatorial analysis. She followed professional tennis and was an avid and excellent player herself. I played recreational tennis once a week with some other guys. In that group, I was average. I did play Ewa once. She easily bested me. Though not a pro player, rest assured that she could beat most guys. The hot topic in pro tennis at the time was the disparity between how much money women earned and how much men earned. One day, down in the dumps, Ewa came into my office. Vitas Gerulaitis, a player on the men's circuit, had argued that one reason for the disparity was that the quality of play was different – that even the 200th ranked man could easily beat Martina Navratilova, the top ranked woman at that time. Ewa said that his statement was true, and it bothered her. How could the top 200 players all be men?

I like problems that can be tackled using probability and statistics, my specialty areas. I showed Ewa that one answer lies in the mathematics of bell curves. Imagine that on some continuous numerical scale, each person's tennis ability could be assigned a number. It is reasonable to assume that tennis ability is normally distributed. Let M denote the mean tennis ability for men and F denote the mean tennis ability for women. It is also reasonable that M should be greater than F – maybe not a whole lot greater, but greater, nevertheless. One reason this would be the case is that, while one can find plenty of examples of women who can beat men, as a group, men are stronger on average than women, so men have a slight group edge – they hit slightly faster shots, on average. Suppose the two bell curves are identically shaped (have the same standard deviation), but that the men's bell curve is centered to the right of the women's (since $M > F$). First note that for any given tennis ability level X, the area to the right of X under the men's curve (call this area A) is greater than the area to the right of X under the women's curve (call this area B). Even if M and F are almost equal, something interesting happens. One can show mathematically that as X increases, A/B approaches infinity. As a consequence, as X increases, the ratio of the number of men who can play at or above level X to the number of women who can play at or above that level approaches infinity. In other words, so long as one group has even the slightest edge, this translates into an increasingly huge monopoly in the number of players at the highest levels of play.

Something that would keep this from happening is if the standard deviation for women's ability exceeded that of men's, which is statistically counter-intuitive. This mathematical argument was enough to convince Ewa.

The other conversation I recall – on a different day when she was down in the dumps – had to do with sexual bias. Before leaving home for work, she happened to catch a few minutes of the Today Show. She was upset with the hostess of the show for being a poor role model for girls. The hostess, who shall go unnamed, was an attractive, well-known celebrity. The guest for the day was some guy who was an "expert" on bias. When the topic of bias in testing came up, the hostess asked for an example of a sexually biased question on the SAT exam. The question the guest used went something like this: *One fifth of a number is 4. How much is half the number?* According to Ewa, the hostess said that she could not work the problem, and so couldn't explain why it was biased, but that the guest could. The guest's justification for the bias was that a higher percentage of boys got the correct answer than girls. Ewa was furious for two reasons. First, because someone as prominent as the hostess would admit to all of humanity she couldn't work such a simple problem, little girls didn't have a good role model in the hostess. And second, the "expert" provided such a mathematically shallow and insufficient justification.

Anyway, I thought you might enjoy these recollections about an old friend who was interested in women's rights and in providing opportunities and role models for girls, but who also was rational.

Sincerely yours,

Mick Norton, Professor of Mathematics

Department of Mathematics

College of Charleston

Charleston, SC 29424

The next day, Mick received a response.

Subject: thank you

Yours was by the far the most interesting response to my column. Your point about bell curves got me thinking about some of the research my organization has done on college admissions and race. With so few black and only slightly more Hispanics scoring in the top decile of the SAT, we may find very few blacks or Hispanics being able to compete for places at the most elite schools. I'm saddened by that, but see no fair way to compensate for it—the Supreme Court decision on the University of Michigan notwithstanding. Given your example of men vs. women, where the absolute numbers of the two groups are roughly equal, it is all the more depressing to think of the same effect occurring among racial groups where whites outnumber blacks and Hispanics by about 4-to-1, making the chances of any particular black or Hispanic competing with those at the right end of the SAT curve very slim indeed. Am I right in this analogy?

Linda Chavez

And Mick responded.

Subject: thank you

Yes. If at a given high level of tennis ability, the ratio of men to women were 200:1 with equal numbers of men and women in the population, the ratio would be 800:1 if men outnumbered women by a ratio of 4:1 in the population.

Again, I enjoyed your column – not just because statistics can be applied, but because it caused me to remember some great discussions with a good friend.

Regards,

Mick Norton

17 COMPARING THIS YEAR'S NUMBER TO LAST

Some courses were more fun to teach than others. For Mick, the graduate course in statistical quality control was one of the most fun courses to teach. He had designed the course for the math department's master's degree program. Actually, he had been the only one who had ever taught it. It was his baby. Part of the fun would be seeing old friends, because the class would take tours of several manufacturing plants, places where he had consulted. This time around, his students would see fuel injectors being made at the Bosch plant in North Charleston, and spandex fiber being made at the Bayer facility in Goose Creek. Not only would the students see products being made, an interesting enough field trip all by itself, they also would see companies actually using the tools of statistical process control (SPC) that the students were learning about in class.

The first class meeting of the course was particularly special to him. Much of it would be used to start the students off on developing a mindset. As a course prerequisite, the students already would have completed some statistics course, somewhere, somehow, and could be counted on to know the basic statistical tools they would need for the course. But beyond the mechanics they had learned and would build on, there was a mindset they likely would not yet have acquired. It was possible to know the standard skills of the discipline but still be an airhead when it came time to apply the skills in the real world. This was a course in which there would be plenty of opportunities for students to have light bulb moments about how exercising common sense and being practical went hand-in-hand with statistical analysis. It was a course in which students would taste the real world.

One of the points he would hammer hard was that to understand statistics, it was essential to understand – really understand – variation. It was all too common for well meaning, otherwise bright and responsible people to react to data that seemed off the mark – to them – by making changes in the system that produced the data. And although the purpose of the changes had been to cause future data to be better, all too often the changes would cause future data to be worse. Making changes based on naïve intuition was a recipe for creating future problems.

He began shaping the mindset by hyping the point that in order to judge whether changes in a system should be made, the amount of variation that was naturally present in the system had to be measured first. Only then could there be a basis for judging what kinds of numbers from future data would be *off the mark* enough to consider changing something. The professor told the story of a mid-level manager at a manufacturing plant who would examine data from recent production runs, see some number that was off the mark – to him – and tell the operators to change some particular thing. Some of the operators, the ones who did the grunge work and who lived with the process and also saw the numbers it produced, recognized – by weight of empirical experience – occasions when the changes weren't called for. But they had to follow orders.

The professor made sure to bring the quincunx with him to the first class meeting. The quincunx, a training device used to help people understand the wisdom of some important quality control principles, looked something like a pachinko machine. Mick would position a funnel at the top of an array of pins and then let students drop balls into the funnel. When a ball came out of the funnel, it would hit a pin and bounce downward either to the left or to the right, where it would hit another pin, bounce downward either to the left or to the right, and so on. It would hit ten pins altogether before falling into one of the columns at the bottom of the device. To help him carry out some of the mindset development activities in class, the professor had numbered the columns from 0 to 10, to reflect how many times the ball would have bounced to the right in order to land in the column. Well, almost. It was possible for a ball to hit a pin's sweet spot, particularly with the first pin, and bounce farther to the left or right than usual. This *random noise* actually made the device a little more interesting. It also meant that Mick had numbered some additional columns, using -2, -1, 11 and 12.

This night, the quincunx would be used to simulate making batches of a polymer. The number on the column in which a ball landed would represent the viscosity of the completed batch. Whether the ball bounced to the left or right when it hit the first pin would indicate whether chemical A was over-dispensed or under-dispensed (the amount of chemical A dispensed could be assumed to never equal, to infinitely many decimal places anyway, the amount called for by the recipe). Bouncing to the left or right at the next pin would indicate whether chemical B was over-dispensed or under-dispensed, and so on. In this way, the quincunx simulated how one variable (viscosity) could behave in reaction to ten inputs, each of which affected viscosity.

Students were told that they would play the role of operators at the chemical plant and that any batch they made with a viscosity of two or less was not viscous enough and would have to be scrapped. With a marker, he drew a bold vertical line between the 2-column and the 3-column to showcase the product specification limit. Each student came up to the front and "made" 20 batches of polymer. Mick recorded on the board the name of the operator and the number of defective batches he or she had made.

This particular summer night proved to be Mick's most memorable first class meeting because Libby, knowing about the common sense principles and activities that would occur, had encouraged their two children to attend. And they did. Andrew had already graduated from Clemson and Susan would be a junior there. Neither was a mathematics or statistics major, but for what would take place before the class took its five-minute break at the midpoint of the first night, that made no difference.

As each student took a turn at dropping 20 balls into the funnel that never moved, the professor provided encouragement and reminders. "Try not to let any balls land to the left of that mark," and "think *to the right*," he would tell them. As usual, several student-operators laughed at the obvious absurdity of the remarks. And at the end of the activity, the summary data were as follows:

Rick	1
Jennifer	1
Tracy	2
Chris	1
Jeff	1
Susan	0
Andrew	4
Clare	2
Arthur	1

"OK, it's time to give merit raises," announced the professor. "Susan, you are obviously very careful in your work. You understand how important it is to make good batches, and you do an excellent job. We're pleased to have you as an employee. You will get a 4% merit raise next year.

"Rick, Jennifer, Chris and Jeff, we are happy you work here also. You do a fine job. Other operators could learn a lot from you. Next year, you will each get a 3% raise. Nice job. Remember to try to be as careful as you can. Scrapped batches really cost us.

"Tracy and Clare, we here at the company believe your hearts are in the right place. Certainly you are not making bad batches intentionally. Perhaps you have forgotten some of the fundamentals that you need to think about as you make batches. Monday morning at 8 am, we want you two to retake the training course that we give to new-hires. Be at Building A-5 at 8 am. The retraining should help you. You will get 1% raises next year.

"Andrew, ... ," said his father, pausing for effect. "Do you know how much it costs us to make a batch and then scrap it?"

"No, but I think I'm about to find out."

"$5000. On top of that, we could have sold four good batches, had you made them. And on top of that, we pay you a salary while you are making bad batches. Certainly what you are doing is not intentional. Actually ... I suspect that you will have a very bright future. Just not with us. You are to stop work now and turn in your uniform. We will, however, pay you for the next 30 days.

"OK everybody. What do you think?"

"What you just did – it's not right," offered Clare.

"Hey, I can defend what I did. I believe in data-driven reasoning. All I did was go with the numbers. Don't you see?"

"But it's all chance," offered Chris. "All of them were operating the exact same process. You can't blame them for numbers that are not their fault."

"Aha! Yes." The professor smiled as he pointed at Chris. "Did anybody measure the variation shown by historical data? No. So nobody can say what the standard deviation for viscosity is when this equipment is used, can they? You need to study the variation present in historical data in order to find out what kinds of viscosities are unusual. If you don't have a basis for knowing what kinds of numbers to consider unusual, then you don't have a reason, based on a number that somebody produces, to suspect that that person may be messing up. You shouldn't be rewarding or penalizing people for numbers that are beyond their control."

Susan seemed to enjoy the idea that she had been rewarded the most, by chance. And Andrew got a kick out of being fired by his father. And Dad enjoyed the fact that his two kids, who disappeared after the class break, had set the extremes in the classroom activity.

EXERCISE FOR THE READER: Given that there are n operators in the exercise, what's the probability that one's two children set the extremes? Make the assumption that no students tie for the most extreme outcomes.

18 PROVING THEOREMS CAN BE TOUGH – TAKE A BREAK – SMILE AT STUDENT EVALUATIONS

Reading the morning paper over a waffle and coffee at Jack's was a great way to start the day. Mick added one scrambled egg on the side, once again living life at the very edge.

After walking to the office, he got down to the first order of business – reading the new emails. Only one warranted a response.

Professor Norton,

I noticed that the grade for your statistics class has already been posted and I have received a B+ for the semester. I have a question about this grade. I am not doubting or making any challenge, and I don't want my intentions to be misconstrued. I am simply curious as to how that was my final grade. I had a 94 on the first exam, and then, if I didn't misunderstand, the second exam grade would be replaced if our work on the final exam relating to that material was higher, which I believe it was. Now, I have not seen my grade for the actual final exam. However, I do believe that I earned higher than an 84, a B+ for your class with the results from the exam and the substitution of Exam 2. I would greatly appreciate a response whenever you may be able to. Thank you very much and I would like to say your class definitely furthered my understanding of math and may lead me toward courses in that field in the future.

Thank you,
Mark Mehier

All emails like this one set in motion a sequence with the same first step. Mick pulled out the grade book, found the course and looked up the student's grades. A wealth of teaching experience over the years had made him aware that students who sent emails like this one wore rose-colored glasses. A glance at the grade book showed that Mehier's needed to be bifocals. The professor tried to write a letter that would convey the news and still be upbeat.

Dear Mr. Mehier,

You replaced the 38 on test 2 with a 91 (very nice), and got a 141 out of 150 on the final (also very nice). This gives you an average of 80.9% for the course, which is in the middle of the B range. What hurt you was four homework assignments not turned in – there were 60 points possible in all. Generally, I don't give "+" grades, except to people who would just barely miss the next higher letter grade. I virtually never give a "+" grade to someone whose average is in the middle of a grade range. However, I did it in your case because your exam grades were so strong and a B did not seem, well, just. Good luck with more math courses. The more math you take, the more options you will have later, even in fields many view as not really related to mathematics.

Regards,
Mick Norton

With the spring semester having just ended and the morning emails having been read, it was a good time to do research. He'd had a revelation on how to attack the Union-Closed Sets Conjecture (UCSC). Maybe it would lead to a proof that that the conjecture was true. As with many problems in discrete math, it was easily stated, easy to understand, didn't require background reading, and compellingly deceptive to the point of making you think you ought to be able to prove it. Mick had spent much time on the problem, on and off for several years, and it had long been obvious that a proof would be difficult or even impossible. He could see no potential application of the conjecture. It was pure math for its own sake. Still, the problem was compelling and he worked on it.

Sometimes he would experience a rare moment of inspiration and a clever approach to a proof would suggest itself. Most of the time, he would be trying to prove the conjecture. At other times, he'd be trying to concoct a counter-example. Years before, he'd similarly spent a lot of time trying to solve Fermat's Last Theorem. A proof had eluded the best math minds for hundreds of years. He was pleased that someone had actually found a proof in his lifetime. It freed up many hours.

The UCSC was easy enough to understand. Suppose there were a finite, positive number of sets making up a collection of sets, that each set in the collection contained a finite, positive, number of elements, and that the union of each two of the sets was, likewise, a set in the collection. The conjecture was that there had to be at least one element that was in at least half of the sets in the collection. Colleague Dinesh Sarvate had made him aware of the conjecture, and the two of them had published a joint paper on a property that would have to be possessed by any counter-example with the minimal number of sets that a counter-example could have.

A paper on a pure math topic was a rarity for Mick. He'd produced plenty of papers answering tough questions in probability and statistics, many of which had ready applications. But when one of those problems was completed, he'd often take a break by returning to a compelling, recreational math problem like the UCSC, one that many would not work on because it was viewed as too tough to crack. For the professor, though, it was like being on a vacation.

Papers that resulted from solving sufficiently difficult problems could be put on a resume and on annual reports. They were good for the professional reputation – and good for the soul. Of course, the time taken to work on a problem that had little expectation of success was time that could have been used instead to work on problems whose solutions would be less awesome but more likely to produce publications. Yet he continually reminded himself that some of his successes had come from trying the road less taken. Perhaps he needed to attend a twelve-step program for mathematicians addicted to trying to solve the toughest math puzzles.

Hi. My name is Mick. I spend too much time trying to solve math problems I probably won't be able to solve.

Hi, Mick.

It was close to noon, and trying to prove the UCSC using this revelationary method of attack had revealed tough questions that would need to be answered if the method were to pay off. It was time to take a break, grab some coffee and check the snail mail.

In every professor's mailbox were those telltale large manila envelopes. *Ah yes,* thought the professor, *the semester has been over for a few days.* The envelopes contained student evaluations. A few colleges and universities had begun doing this online, but the College wasn't there yet. He opened one of his envelopes and started to read the written comments.

> This is the best professor I've ever had in my life. Has a knack for knowing just the right words to explain a concept. Enjoyed the class immensely. Great sense of humor, and very sensitive to students' feelings.

What a great first evaluation, Mick thought. Would the rest of the students in that class say similar things? As he turned the page get to the next student's comments, he happened to glance at the top of the page, where it said *MATH 502 Instructor: Anna Calini.* He put the evaluations in Anna's mailbox, where they belonged. Bummer.

Walking across campus later that day, he exchanged hellos with music professor Doug Ashley and they stopped to chat. The conversation turned to the timely topic of student evaluations and what they measured. Professors who had been around evaluations for any length of time usually agreed that students didn't have the background to determine if they had received a good course. However, everyone agreed that the one sure thing evaluations measured was whether the students were happy. "Some students should not be allowed to evaluate professors," Doug opined. He said that he had once received an interesting evaluation, one that he would never forget:

> Your course is to hard. And I still do not like box music.

Mick proceeded to the lounge to join his math colleagues for lunch. Kathy Thom was already there. She was one of the regulars in the lounge at lunchtime. Her dad's experiences as a soldier in WWII had prompted Kathy to become very well read about WWII military operations. She told Mick that she had put something to read in his mailbox. It was the chronicle of her father's experiences in WWII.

He read Don Johnston's chronicle that afternoon and evening. It began with his military training in the states, moved to his combat experiences as the Allies advanced across Europe, and ended with his experiences after returning to the states at the end of the war. Mick found it to be as riveting as *Marine Sniper*, the book about Carlos Hathcock's Vietnam experiences. Suffice it to say that Don Johnston had been in a succession of heavy-duty engagements. He had seen many die and could easily have been killed himself. His writing was detailed.

Kathy was instrumental in creating the electronic version of the chronicle. She said that there were some war experiences that her dad would never talk about, that she had found out about them only by reading the chronicle while transcribing. It was fortunate that he had kept a record. Mick asked Kathy for her dad's email address and wrote.

Dear Mr. Johnston,

Kathy let me read a copy of *Over Here*. I'm glad I got to read what you wrote. Today, our armed forces are all-volunteer, which means that most people don't have family members in, or even know anyone in, the military. And so, many do not have any concept of what the military is for or of why anyone would ever want to serve in it. We live in weird era when duty, honor, country and sacrifice are politically incorrect concepts to many. As a kid, I had no immediate family members who were in the armed forces who could serve as role models. I got drafted out of grad school and wound up in Vietnam. Kathy and I are in academics, where many faculty members are *progressives,* some of whom say outrageous, clueless things. Maybe if I hadn't served, I'd be saying nutty things, too.

One thing I can say is thanks. The world benefited from what you did and from the ultimate sacrifice of many you knew. I'm glad you came through it. I'm also glad you were able to record your experiences. All the best,
Mick Norton

19 PROBABILITY AND ISLAMIC TERRORISTS

The morning newspaper carried an AP article about some federal government agency that was trying to estimate the chances that Islamic terrorists would explode a nuclear device within the U. S. in the next five years. This was just the kind of example Mick could use to illustrate a concept that was to take up a few minutes of the next day's freshman level introductory statistics course. He had planned to use a different example that also illustrated the point perfectly. But in light of the newspaper article, the nuclear device example would be timelier.

So in class the next day, he stated the question: *What is the probability that Islamic terrorists will explode a nuclear device within the U. S. in the next five years?* He remarked that this was an example of a probability that could never be known nor even estimated by simulation. Although the event would either happen or it wouldn't, you couldn't say that it had a 50-50 chance of happening (as in getting a head on a coin toss). Further, you couldn't estimate the probability by examining the outcomes of a large number of replicated 5-year trials. The question was similar to the mainstay question that had been used in statistics classes during the Cold War to illustrate the same point, one that Mick also had used for years – *What is the probability that the U. S. and Soviet Union will ever destroy each other in a nuclear war?*

After class, Jonathan, one of the students in the class, jogged to catch up with Mick as the latter walked back to his office.

"Hi, professor," offered Jonathan. "I've been thinking about something you said. Your example about the nuclear device – it's not a good example to use."

"Why is that?'

"There might be some Muslim students in our class, and they might be offended."

"Gee, I sure wouldn't want to offend anyone." The tone conveyed thoughtfulness and concern. "I tried to be careful in my choice of words. Remember that I didn't say Muslims. I said Islamic *terrorists*. I tried to be careful about that."

The student chewed on this for a few seconds and made a reasoned pronouncement.

"Yeahhhh, …, you're probably OK."

In the 1970s and 1980s, no student had ever expressed concern that there might be a student with Russian ancestry who might be offended. *What a relief,* Mick joked to himself. *The student had judged that I was 'probably OK.'*

After class, he felt a need to unwind. It had been a long day of teaching and committee meetings. He had been working on a statistics research question and wanted to get back to it. But the right mood would be needed for that. So he took off across campus and started walking down Montagu Street. Like much of downtown Charleston, Montagu Street had houses that were pleasing to the eye. It had been some time since he'd had occasion to walk down the street. As he walked, he looked for changes that had occurred. The shrubs of a couple of the houses were a little classier, a wrought iron fence had been recently painted, and a couple of houses had changed colors – a very rare event for downtown Charleston. Not much else had changed since he had lived in the neighborhood. He turned left at Rutledge Avenue. Most of the houses there were not only pleasing to the eye, but also grand and huge. When he used to live in the area, he'd paid attention to sales prices. *What do they go for now?* he wondered. *A huge number of grand, for sure.*

After two cross-streets, he came to Colonial Lake, a man-made trapezoidal body of water bounded by four streets. Three of the four streets had a backdrop of gorgeous houses. To the west, across Ashley Avenue, the street that bounded the fourth side, there was Moultrie Playground. The north end of the playground had shade trees, picnic tables, and playground equipment for small children. At the south end of the playground was a sandlot ball diamond. Mick had fond memories from when his kids were small, of the many times he had taken them to the ball diamond with gloves, bats and balls so that they could practice hitting and throwing. Not far beyond the park and ball diamond was the Ashley River.

Having arrived at the eastern edge of Colonial Lake, the professor plopped down on a bench in the shade. It was relatively quiet, and the breeze coming off the Ashley River made it an ideal place to ponder the statistics question. During a half hour of good thinking, a few people arrived to do their constitutional laps on the sidewalk that went around the lake. There was a male jogger dressed to show off his physique, a pair of older women engaged in a relaxed amble and chat, and a woman in her thirties briskly walking a black lab. Mick noticed each one of them precisely once per lap – at the moment each passed right in front of his bench.

On the opposite side of the lake a young woman arrived to become a second lap walker. She looked familiar in the distance. Within a few minutes she neared Mick's bench and they smiled at each other in recognition.

"Hi Clare."

"Hi, Dr. Norton."

Clare was a student in one of his classes that semester. She was not a math major, but was doing A work, better than almost all of the math majors in the class. What Mick liked most about Clare was her work ethic. She was an ideal example of someone who burned the midnight oil studying, a fact he'd become aware of during conversations with her at his office. She was a regular visitor during office hours, always prepared, always bringing discerning questions, always able to figure out the puzzle of the moment when given the slightest of a hint. In short, Clare was a bright, serious student.

She also was very much at ease making small talk – which could be about anything. She was not intimidated by people with degrees. Neither was she full of herself. She was always relaxed and bubbly. *The world needed more Clares*, he thought.

Clare sat down on the bench and they chatted for 20 minutes or so – about current events, books they'd read, and the current presidential race. And they shared some laughs. Ultimately they both got up and went their separate ways. As he walked, he thought about joggers, walkers and bench sitters. *How many people had sat on these benches or gone round the lake over the years?* He mused about neighborhood traditions as he made his way back to campus. And he mused about Clare. Watching students blossom – students like Clare, who worked hard to be high achievers – was one of those special rewards of teaching. *And maybe Jonathan took a small step today.*

20 PICK A NUMBER, ANY NUMBER

The occasion was Andrew's 27th birthday. Susan and her fiancé Ryan, Andrew and his gal Leslie, Libby and Mick all met up at a Mexican restaurant on Meeting Street. Ryan had proposed to Susan not long before, so that topic added to the light, festive conversation as the group ate at an outdoor table. Joining the group in food and drink was a group of perseverant flies.

For the occasion, Susan had bought a cake with white icing on which she had written *Happy Birthday Andrew* using chocolate icing. One could easily read the words when she had put them on the cake. Looking down at the circular top surface of the cake one could see *Happy* near the top of the circle, *Birthday* across the middle of the circle, and *Andrew* near the bottom of the circle. The cake was transported to the restaurant in style, inside a transparent plastic cake container that sat on the back seat of the car. It was a bright, sunny day, no doubt a good one for greenhouses. By the time the cake arrived at the restaurant, the chocolate had reorganized itself, creating a chocolate face with two white eyes, a white mouth, and numerous chocolate warts that stuck out from the circumference of the face. With seemingly intentional Halloween-like artistry, the eyes were narrowed, slanted in anger. The mouth was ghoulishly contorted. The face screamed agony and malevolence. The professor was sure he was not the only one thinking of Edvard Munch's painting "The Scream," warts and circular face notwithstanding. The cake only added to the festivities and laughter. It was easy for people to get a case of the sillies.

Andrew wisecracked, "Maybe I could sell this cake on E-Bay. It's not an image of Jesus on a cabinet door, but who knows? It might sell."

Estimates were suggested for how much it might fetch. Plenty of photos were taken of everyone gathered together at the table, and plenty of the cake.

Digging in her purse, Susan retrieved and positioned one lone candle on the cake. It had a thick, white center-post shooting up through a thicker, green, five-point star. The wick was at the top of the center-post.

"Of course, Andrew, you know why this particular star-candle is used only for 27th birthdays," offered Dad. Andrew looked at his father with a little smile that said *here we go*.

"No, why?"

"You get five years for each point on the star, one year for the white part above the star and one more year for the white part below the star."

Susan and Libby fought hard to keep from rolling their eyes, but went along.

"Dad, there is no such significance to this star," responded Andrew, who may have been ready to have the cake ceremony begin.

"Dad, it's just a candle I happened to pick," proffered Susan.

"Aha, you don't believe me. I know all about these birthday traditions."

Ryan entered the fray. "I thought this candle was used for the 26th birthday. The white post only counts once."

"You're thinking of the one with the red star," responded the professor.

"Yes, that's right." Ryan smiled and nodded, feigning recollection.

At this point Susan produced a lighter. It was cylindrical, had a shell of orange plastic, and was inexpensive. Susan did not smoke, and Mick suspected that she'd bought it at a gas station on the way to this event.

"And Andrew, you know the history of using only orange lighters for lighting candles for 27th birthdays," prompted the professor.

"Dad," protested Susan, "c'mon now. This is just a lighter." It was clear that Susan was ready to begin the ceremony of lighting the candle.

"Susan, you may not believe me, but I know these things. I kid you not. One day you will have absorbed all of this cultural knowledge, too."

Acting deadly serious was part of the fun for the professor when he got silly. The delivery of his last sentence made it clear that Dad was done. He was ready for the candle lighting as well. *Maybe I did go a little overboard?* he wondered.

The cake was delicious. Susan had made a good choice. And of course the small talk and good times continued as they ate. At some point, however, everyone was sharing in the conversation except Andrew, who had picked up the lighter and was studying it. The plastic shell was translucent. Perhaps he was studying the insides.

He reentered the conversation with an announcement. "There's a 27 on the bottom of the lighter. You can hardly see it, but it's there." The lighter passed from person to person so that people could see the bottom for themselves.

Were Andrew and everyone else getting silly now, doing so with thespian honesty? When it made its way to Mick, he squinted to examine the bottom of the lighter. The professor almost needed a magnifying glass to see it, but it was there. A very tiny 27 was stamped into the bottom.

"And you guys didn't believe me. Maybe next time you'll listen."

21 YOU CAN'T PLEASE EVERYBODY

It was the middle of the fall semester and Mick was teaching MATH 104, again. One of the homework problems that students wanted to see him work was about a hypothetical company that chose a committee of employees from its workforce. The size of the workforce and committee were given, as was the number of minorities in the workforce. After pointing out that no minorities wound up on the committee, the problem asked the students to find the probability that such a committee would result from a process that chose people at random. The answer given in the back of the book was .006. Also asked was whether the probability indicated that there was bias in the selection process. The back of the book noted that the probability was small, so that the answer was YES.

Problems like this one were there to help set the stage for an upcoming topic, the drawing of inferences based on data. The professor reviewed the mechanics of what was needed to obtain the probability, then found the total possible number of committees that could be formed without minorities, found the number of possible committees that could be formed without restrictions, and divided the former by the latter to obtain the .006. He then observed that YES, by its simplicity, was potentially misleading, in that one could not draw a sacred line of demarcation and say that if the probability was smaller than that threshold, that the company must have used a biased process, whereas if was not smaller, then the company must not have used a biased process. He observed that the small probability in this problem was suggestive and that it did, in fact, meet the rule of thumb the class had been using for declaring an event to be unusual (a probability of .05 or less). But he also cautioned that the small probability was not absolute mathematical proof that the selection method was biased.

"Is this something we'll see when we get to inference?" asked Ryan. Ryan was one of the really sharp students in the class.

"Yes it is. We will soon explore the ins and outs of drawing inferences from data. One of the things we'll learn about is how often to expect that the conclusions we reach are wrong."

"I like the book's answer. That probability is too small," announced another student – Lena. Her pronouncement sounded defiant. Lena was a bright student who never said much in class. The professor saw a teaching opportunity.

"Does the answer *have* to be *Yes?*" he asked. "The probability is small, but is it *absolute* proof? Couldn't such an outcome have come from a fair process?"

"Well, yes," replied Lena. "But I will go with the book's answer."

"You won't want to phrase it that way on a test. Actually, we'll be talking pretty soon about how to word such responses. All I'm saying now is that YES is a very simplistic answer, one we will elaborate on, as you will see in a few days. At this point, all I want you to be able to recognize is that the evidence is suggestive, but not absolute proof."

"Well, yes."

Mick then explained about having expert witnessed in court in discrimination cases, observing that small probabilities provided circumstantial evidence, but almost always in conjunction with other evidence. He then described one of his favorite cases, one that was an exception to the rule. It was an age discrimination suit, the only case he'd ever had in which the company was willing to settle out of court solely because of a small probability. The company had 144 employees, 15 of whom were in the 60-or-older age group. Nine of the 144 had been terminated, the youngest of whom was 60. The company maintained (A) that the nine were terminated only because of poor job performance and (B) that there was no statistical relationship between age and performance. He had worked on the plaintiff side in that case.

The professor asked his class to compute the probability that a simple random sample of nine employees would consist entirely of employees in the 60-or-older age category. The students computed the answer using the same method that was used to answer the homework problem – they divided the number of ways that nine 60-or-olders could be chosen by the number of ways nine people could be chosen without restriction. The resulting probability was .000000000088.

Putting the finishing touches on the impromptu exercise, Mick continued. "In connection with the case, I submitted a report containing my statistical analysis. There were other statistical issues besides the problem you just worked through. In a deposition, attorneys from the company grilled me for hours about the report. Mostly, though, they grilled me about the analysis that produced .000000000088. The day after the deposition, the plaintiffs' attorney called to tell me that the case was over. Solely because that one probability was so "microscopic," the company settled out of court.

"Typically, when statistics is used in a case, it is just one of a number of pieces of evidence. It is really rare for a case to hinge on just a single probability."

"Could a company argue that advancing age affects job performance?" asked Ryan.

"Sure," Mick responded. "That issue would be a natural one to crop up. Here though, the company had said that age had no connection to job performance, so that was off the table."

To his recollection, Lena had never said anything in class before and he decided to encourage a bright student. When class was over, the professor walked over to her and commented that she had been eloquent, analytical and passionate on the topic that day. Her response was testy.

"You are insensitive. You know, there are very few black students at this college – not enough. You presented homicide data several weeks ago (see Table I). We have several black students in this class and I'm sure some of them would have been offended. You can't use data like that in a class – maybe in a psychology class you could, but not in a statistics class. Your comment just now on the text's answer is another example of your insensitivity." She turned and walked out in a huff.

The only student left in the room was Harvey, who had seen Lena leave and saw the professor standing there dumbstruck. Harvey shared an observation. "She's really way out there, isn't she?"

Mick walked back to his office in sad spirits. More and more students seemed to come to college indoctrinated, wearing politically correct blinders so big that they were unwilling to examine certain areas of real life. Not only was discussing certain topics taboo (unless you were spouting a PC view). Also taboo was examining real data that might shed light and understanding on the topic. More and more professors in the mold of Arlene Credenda were being hired, and shaping how their students viewed the world. And many of their students would become K-12 teachers who would shape how their students would view the world. *And so on*, he pondered.

Flashbacks flashed as he walked. He and Libby were eating at a restaurant one Friday evening when young woman approached the table.

"Aren't you Dr. Norton?"

She introduced herself and said that she had been one of his students a few semesters earlier. She identified the course and the semester. She had majored in social science and had since graduated. She did social work for some government agency, a job she said she loved.

"I have to confess," responded the professor, "that I don't remember your face, although I do remember your name. And I do have fond memories of that class. It was a good group of students. I do remember that."

"I never cared much for math," she said. "But that course was fun. Remember the homicide data? That example helped me understand some of the things that I have to deal with at work."

Mick also recalled a black, female student who, two years earlier, had told him in class that the data set had been both revealing and interesting. Just recently, in fact, he'd run into her at a food court and she'd invited him to sit down and join her for lunch. Being asked was an upper because she presented an interesting combination – she had been a non-math major, she had frequently professed her belief that she had little math ability, the course was a requirement for her, she had done well in the class, and she enjoyed it. Of course, the invitation was not much of a surprise, given that Kimberly had one of those bubbly, likable personalities that money couldn't buy. The world needed more Kimberlys. Over lunch she proudly announced that she had just been accepted into a graduate program at a top-tier university. *Way to go, Kimberly.*

To Mick, one major goal in an introductory statistics class was to have students see how data could help them understand real life. It made no sense to have a statistics class in which people would learn how to interpret data, but be sensitive to the point of disconnection from the real world. He had used the homicide data many times. The computations and discussion used up about 15 minutes of one day in a one-semester course. It was real life, had a local flavor, was a fascinating and revealing data set, and was a natural for showing students how to compute conditional probabilities from a table. And now he had a complaint.

And what of the book's homework problem – the one that had set Lena off? Students had wanted to see him work it, and he used the book's overly simple interpretation to help set the stage for the next chapter in the book. What was he supposed to do? And, he thought, why could a psychology class consider such data, but not a statistics class? The whole thing was nuts. Differences between students was one of the things that made teaching interesting – and challenging. And sometimes aggravating.

22 THIS WILL STAND YOU ON YOUR HEAD

It was a Tuesday, a non-teaching day for Mick, and there were no committee meetings. The next day, he would be covering a topic in mathematical statistics that he'd never taught before. So Tuesday afternoon would be devoted to making the next day's lesson plan. But this was Tuesday morning, and the time for morning rituals. Ritual number one was checking new emails. A friend suggested that he take a look at a website that appeared to contain an email he had written.[viii]

NEED TO LOSE ABOUT 8 POUNDS IN ONE WEEK! AH! Lol

Cara 15 February 2003 16:16 [in reply to someone who was looking for a good way to lose weight quickly]

Hi! I compete at 60 kg, but normally weigh between 63-4. I often have to diet down really quickly. The amount you want to lose can be achieved – if you use the Mayo diet. This is a specially designed diet aimed at fast weight loss whilst maintaining maximum strength. My coach Ron Reeves (best coach in the world) can be contacted at his gym:
Rons Gym
Periwinkle Close
Milton Regis
Sittingbourne

Kent

England

He will be able to give you details of the mayo diet. It works on chemicals. You have to like eggs, steak and olives!! It's not very pleasant, but really works – I've never weighed in heavy!! You can only go on the diet for a two-week period.

Don't do saunas/sunbeds – you lose too much strength!

A glass of dry white wine the night before helps!

Also, standing on your head for a few minutes before getting on the scales (my training partner lost 0.7 of a kilo doing this!)

Hope this helps – Cara

JohnRov   17 February 2003 18:41

I hope this statement [about standing on your head] is a joke…

wrstlingoo 17 February 2003 22:14

It's true, Olympic wrestlers do it.

JohnRov   18 February 2003 14:57

OK, then. What does it do? Provide some evidence for your assertion and explain how standing on your head actually decreases the mass of the human body.

JohnRov   18 February 2003 19:00

This got me to thinking so I did some research. I contacted a professor who had done research into this very phenomenon. The conclusion is that people don't actually lose weight, so the method won't work for actual weight loss. It does, however, sometimes help make weight if someone is missing by a fraction of a pound, 1/20 according to Professor Norton. I will let his explanation provide all the details. This is his reply:

Dear Mr. Rovnan,

I am a former wrestling official, as well as a mathematician with specialty in probability and statistics. Beginning in the mid-1980s, wrestlers who still needed to lose a fraction of an ounce when the weigh-in period was about to expire began standing on their heads for a minute or so, then would get back on their feet, step on the scale, and surprisingly often, qualify for their desired weight class. The practice is now banned in international, college and high school wrestling. Every official who has supervised weigh-ins has seen the headstand apparently cause weight to be lost. I accepted what I was seeing for many years, but when the topic would ever come up in conversation, people outside the wrestling community would say "That's crazy," or "Mass doesn't go away," and then ask "Why on earth would it work?" Even a few coaches questioned whether it worked.

No one had ever collected data to confirm that it worked. So I brought my academic specialty to bear on my hobby. I visited practice sessions of local coaches Paul Spence (Bishop England HS) and Martin Williams (James Island HS), who were kind enough to let me weigh wrestlers after their warm-up drills. Twenty-three wrestlers were weighed. Each wrestler was weighed twice, with a period of 1 minute, 45 seconds in between. Method I, where the wrestler did a headstand between weighs, was compared to Method II, where the same wrestler simply stood idle for the 1 minute, 45 second period. There was a third method, too. In a nutshell, Method I produced a significantly greater weight loss than Method II. There was a roughly one-twentieth of a pound weight loss attributable to the headstand, regardless of the weight of the wrestler.

Also, if the wrestler were allowed to remain on the scale, after about three minutes, it was typical for the scale arm to begin inching upward, indicating that the mysterious weight that was lost was being regained.

An explanation comes from physics and anatomy. When you do a headstand, blood drains into your head. Then when you get back on your feet, the blood drains back out. An anatomist friend says that spinal fluid and interstitial fluid do this also. When small amounts of these fluids are falling, (more importantly, accelerating), they don't contribute to weight until they land. When all excess fluids have drained out, the weight "lost" reappears. As a parallel example, put handles on the ceiling of an elevator. Grab onto the handles and weigh the elevator. You get your weight plus the elevator's. Then weigh the elevator again – after you let go of the handles, but before you land.

The standard deviation was larger than $1/20$th of a pound, indicating a lot of variation from body to body. In any case, a small amount on the order of $1/20$th of a pound often meant the difference between qualifying and not qualifying for some wrestlers

Out of curiosity, where did you see that I had done some research on this?

Sincerely yours,

Mick Norton

Neat, thought the professor. He figured he was responding to just one guy. He had no problems, though, with the guy making it available to anyone with the inclination to Google *weight loss* and *headstand*. The experiment had gotten other notoriety, too. A statistics book[ix] by Peck, Olsen and Devore used his experiment as an attention-grabber example of simple linear regression. Mick had even included the data set in the book he had written on statistical process control (SPC) – in the section on scatter plots. Of course, the weight loss data didn't have anything to do with how to collect or analyze data from a product manufacturing or service industry setting, the usual home turf of SPC. However, scatter plots were used a lot in SPC, and Mick thought the weight loss scatter plots were interesting and, most importantly, it was his book. So what the heck.

Occasionally, some of his math department colleagues would encourage him, with a wink, to open a weight management center. Another informed Mick that he had a friend who was trying to lose weight and that he'd told the friend to see him. Among the gifts Mick would receive at his department retirement party was a plaque, engraved with the message "Thank You, Dr. Mick Norton, for Over 33 Years of Wrestling With College of Charleston Students." Also engraved on the plaque was a picture of a wrestler doing a headstand. The drawing on the plaque was required to be unrealistic in one respect, of course. The wrestler was wearing a uniform. An actual wrestler who needed to do a headstand would have just failed to make weight wearing no clothes.

One physics major at the College, after hearing one of his professors mention Mick's experiment during a class, built a huge teeter-totter apparatus out of two-by-fours. At the end of one arm of the teeter-totter he hung a "bucket" filled with water. The bucket, which he'd made out of large-diameter PVC pipe, was about four feet tall. Submerged just below the surface of the water was a small bag of sand, hung from the arm using a short length of string. At the end of the other arm of the teeter-totter, he hung some weights, just enough to balance the teeter-totter. He built the device for a research project that he presented at a meeting of the South Carolina Academy of Science. During the demonstration part of the presentation, he cut the string with a pair of scissors. As the bag of sand fell through the water, that arm of the teeter-totter rose. When the bag reached the bottom of the bucket, the teeter-totter rebalanced. The water played the role of slowing the fall so that the reaction of the teeter-totter would be obvious to the naked eye.

For sure, he'd gotten more mileage from his weight loss data set than he probably had from a few of the deep, math-for-math-sake articles he'd published, the ones that were read by maybe six people on the face of the earth.

23 ADAPT AND OVERCOME

The email that morning contained the usual mix. There were notices of committee meetings, an alert about a lost magnetic parking card, an announcement from Human Resources aimed at faculty members who were about to retire, an announcement that the lost magnetic parking card had been found, and a lot of other announcements that didn't affect him. It usually didn't take long to skim and cull. For example, Mick wouldn't be responding to the internal announcement about the electrician's job that had just become available at the College.

Actually, no emails required a response from him this morning. But there was one that made him smile. It had been sent to all faculty and staff in response to a nutty email that had been sent to all faculty and staff by a colleague from a different department. Mick smiled because the email reminded him that he never received the first email or, in fact, any emails any more from the nutty professor. The guy had been the catalyst for Mick learning how to use his email filter at work.

Email had been intended to be a timesaving convenience. And in many respects it was. But for years now, senders had not needed to judge whether a message was important enough to make hard copies, put them in envelopes, address the envelopes, and snail-mail them. The consequence was that, all too often, every living being who worked at the College was a recipient of drivel that would emanate from a single *send*. Didn't those folks have enough real work to keep them busy? Recipients then were forced to take time to determine what was good and what was chaff.

The College had done one really good thing. Although some time-wasters still got through, it had taken steps to lessen the number of probable-spam emails sent by external senders.

Mick addressed his own internal sender problem by filtering out emails from three colleagues whose messages had always been on predictable, philosophical or political themes. Before learning how to filter, he had always read email titles to decide whether to delete, and it was not lost on him that he had deleted every single email that those three had ever sent. The catalyst that had caused Mick to learn about filtering was a psychology professor whose opinions, expressed in the volumes of emails he sent, were rather, ah, psychedelic.

The professor felt impish pleasure, but also guilt, in having filtered out three internal senders. There were many hundreds of wonderful faculty and staff at the College. He rationalized that filtering out only three didn't seem like a very big deal.

After finishing with email, he navigated to his favorite internet sites. One of these accessed job listings in statistics and statistical process control. Because he taught these subjects, he had checked these sites for years, keeping an eye on the kinds of jobs that were available to his students – and maybe to him, if he had the yen when he retired.

One site asked for key words. After he entered *statistics, statistical process control*, and *quality*, a list of available jobs popped up. Two of the listings were not routine. The first was from a business that needed someone to help ensure environmental quality. When he clicked on the link, he discovered that a bar needed a bouncer.

The second opening was for a quality engineer. That was a standard title. What was interesting was that the company doing the hiring was the Batesville Casket Company in Batesville, Mississippi. Naturally, every manufacturer aspired to have a process of high quality and to make a high quality product. *Customer complaints would never be an issue for this job*, Mick quipped to himself. Nor would customer satisfaction surveys. If everyone in Batesville got buried in a Batesville casket, the late W. Edwards Deming, one of the world's giants in the quality management movement, surely would have been happy. Deming had believed in sole source suppliers. Heck, Deming might be in one himself. A bachelor's degree and five years of experience in "wood manufacturing quality" were desired. Experience in the casket business was not.

It was an hour before class, the first of two MATH 104 classes he would teach that morning. The afternoon would be devoted to research. He began looking over his class notes. Professors spent a lot of time thinking about how to present topics, even about the exact words to use to get a particular idea across. He had taught the course from the current text several times. The first time had been the toughest, of course. Every book had its own way of coming at certain topics. It was the professor's job to see how the book approached things and to be in harmony with the book, while also taking advantage of great examples and approaches gained from personal experience. Personal experience came from having taught out of other texts, from reading about teaching ideas in journals and websites, from ideas shared by other professors, and from practicing the craft itself. Professors also experimented with their own approaches.

To illustrate some of the concepts he taught, Mick would use consulting projects he'd worked on with manufacturing companies. He tried to choose ones that were attention grabbers. Besides bringing a real-world flavor into the classroom, he believed that if he showed that he had actually done this stuff himself, it would help the students believe that they could do it, too.

One of the topics to be covered that day was Chebyshev's Inequality, a formula that provided an upper bound [a number that could not be exceeded] for the probability that a particular kind of event would occur. [There is no need here to be descriptive about the kind of event.] Because the upper bound was a worst-case scenario, it was typically so much greater than the actual probability that it was not useful, in practical terms, as a ballpark estimate of the probability. However, one of Mick's favorite examples was of a practical application of Chebyshev's Inequality. It had come from a consulting gig. A company made a product (he was not allowed to identify the product) that could emit radiation leakage after installation. Measurements of the amount of leakage from twenty installed units had been obtained. Every measurement was well under the radiation limit mandated by a federal standard. The company needed a worst-case estimate of the percentage of all of the units it made that would not meet the standard. The professor had solved the company's problem using Chebyshev's Inequality.

Showtime, he thought to himself as he finished his gear-up for class. *A cup of coffee and we're ready to roll.*

Later, as his morning class drew to a close, he felt one of those inner glows that teachers sometimes feel. He had been really clear with his presentation, he thought. And he felt he had seen recognition in the students' faces.

It had been about a month since Lena Hardaport's after-class outburst. Students were filing out, and Lena was still packing up. He got her attention as she started to leave.

"Have you thought any more about what you said after class a few weeks ago?" he asked.

Down deep, the professor was hoping for an apology. The odds might have been slim that he'd get one, but it was possible. Now that the focus of the course had turned to how to make reasoned decisions based on data, would having seen numerous real-life applications in recent weeks have helped her come to understand that statistics, as a subject, helped people understand how the world worked?

"I noticed that you haven't used the homicide data any more since I complained."

"You think that's because of you?"

"Perhaps. Why do you teach freshmen anyway? I looked you up. You teach plenty of very advanced courses. You don't need to teach freshmen."

"Everybody here teaches freshmen."

She turned abruptly and left. *This is most definitely a case of Apache fog,* he mused. *It'd be nice if, every now and then, crusaders would go through the exercise of reweighing their core beliefs.*

Harvey was almost always the last student out of the room. He hadn't left yet and he had seen this exchange, too. He and the professor happened to make eye contact. Harvey didn't say anything this time, but the expression on his face spoke volumes. *She really is way out there, isn't she?*

Mick returned to his office to ponder it all. He recalled another student, not Jonathan, who had once suggested that he not use an example. Mick had used a morning's front-page, feature newspaper story[x] to introduce the topic of probability. There was a bill in the U.S. Senate to allocate 100 million dollars to investigate a "plague of crime in rural America." The article included data that was supposed to give credence to the existence of a plague. Yet to anyone trained in interpreting data, it was clear that none of the data did. His favorite excerpt was a comparison of the rate of increase of homicides in Montana from 1989 to 1990 to the rate of increase in Los Angeles. Montana homicides had gone from 23 to 30, a 30% increase, whereas LA had jumped 106 homicides, from 877 to 983, but only a 12% increase. Mick told the students that they would soon learn to think in terms of comparing probabilities to assess unusualness rather than comparing percent increase because, when dealing with small numbers, small changes could translate into large, misleading percentage increases. For example, in a region that averaged 1 homicide per year, it would not be unusual to have a year with one homicide followed by a year with two, a 100% increase that was no big deal. The students would soon learn, he added, that if a region averaged 23 homicides per year over the course of many years, having 30 or more homicides in a year was roughly a *nine time in 100 year* event, not particularly unusual, whereas for a region that averaged 877 homicides per year, having 983 or more in a year was roughly a *two time in 10,000 year event.*

He'd used the Montana/LA comparison for years because it was so outrageous. Yet two years earlier, a well-meaning student had suggested to him after class that murder was a topic to avoid because there might be a student in class who knew someone who had been murdered.

On the plus side, comments like these were rare. On the minus side, they never used to occur. Never. Hyper-sensitivity was slowly becoming a social benchmark. He skimmed the exam he had made the day before. The students would take this test on Friday. He began looking for signs of his own insensitivity, wondering if maybe over time he had grown insensitive without recognizing it. It was introspection time.

Some of the exam questions were settingless and purely computational. Several others were word problems that dealt with data from a product manufacturing environment. A few questions had been stolen from the text – even-numbered homework problems he had never assigned. *No issues there*, he thought to himself.

He smiled when he read questions 6 and 7, two questions he had written himself.

#6. For a woman who gives birth at age 35, the probability that the baby will have Down syndrome is approximately .002. Estimate the probability that at least 7 babies will have Down syndrome in a study involving 2000 babies born to 35-year-old women.

#7. It is well known that adult Martian weight has a mean of 60 pounds and a standard deviation of 5.4 pounds. If the population of adult Martians consists of 1100 individuals, estimate the probability that a random sample (without replacement) of 150 adults will have a mean weight of more than 60.5 pounds.

Mick had used questions like #6 and #7 many times. The one-in-just-under-500 chance for Down syndrome was a probability he remembered from a discussion that Libby and he had had with her ob/gyn when Libby was 35 and pregnant. There was a good chance that over the years, one of his students might have had a Down syndrome sibling. But that couldn't be a reason to not use the example. In fact, the typical introductory stat text was chock full of examples and homework problems drawn from medicine, biology and healthcare.

For years, he had enjoyed the convenience of making up questions about Martians. They didn't take much time to concoct because he could test concepts using made-up numbers. Once, a student fighting hard to work her way up to a B in a course had told him that those Martians were giving her fits on tests. Another had once asked Mick how he had come to know so much about Martians.

Maybe question 7 was insensitive to overweight Martians, he wisecracked to himself. This observation rendered the official conclusion to the self-audit.

It was Thursday, and he looked forward to having dinner boost his spirits. One night each week, Libby and Mick tried to have dinner with their two grown children. This was one of those nights. Sometimes Andrew would bring his girlfriend. Sometimes Susan would bring her boyfriend. The get-togethers made for good small talk over good food.

That night, it was just the four of them. They converged independently on the local Cracker Barrel restaurant at 6:30. Andrew ordered breakfast, something he often did at night, while the other three ordered more typical evening fare. In due course the server brought the four meals and asked if anything else was needed. There was momentary silence as they all scanned their plates. "I think we're good," Mick said, and the server left.

Susan reached for a biscuit and realized that the table had no butter. The waitress was already out of sight and the biscuits were hot. Dad the problem solver rose without speaking and headed toward the archway leading to the kitchen. Once at the archway, he looked into what he could see of the kitchen, and then turned to look at the waitresses who were in the main part of the restaurant. The right waitress was not in sight anywhere. Glancing back toward the kitchen, he saw a woman of obvious authority standing on the customer side of the archway. If Oprah Winfrey were in an action movie and needed a stunt-double, this woman would have been perfect. The resemblance was uncanny. She was entering information onto a computer screen. She and a waitress, not the right one, were preoccupied with the screen. Neither noticed him.

His view into the kitchen was limited. So he took a step through the archway, giving himself a good look into the kitchen. Lots of staff and servers were there, but not the right server. He turned, intending to go back to the table and hoping to spot the server somewhere along the way. In the middle of the turn, right there on the counter in front of him, was a huge open-top box of little plastic tubs of butter – the kind with the peel-off top and enough butter for one biscuit. Seizing the opportunity, Mick took six out of the box and started to return to the table.

"What are YOU doing in here?" the woman of authority asked. She had a twinkle in her eye and a smile on her face. The butter tubs in his hand were visible for all to see. Everyone in the kitchen was smiling and watching.

The professor grinned broadly in return. Caught. When a mother caught her child at the cookie jar with cookies in his hand right before dinner, it wouldn't do the child any good to try to explain it all away by saying *I was getting some cookies.*

"Getting butter," Mick said. And in the spirit of the child caught red-handed, he felt a need to explain further. "Actually it was not my intention to come back here to get the butter myself. I was looking for our waitress to ask her to get butter. But you see, there's this whole box of these right here."

The staff was laughing and smiling.

"Yes, I can see that. But what are you doing in HERE?" the woman of authority asked. Now her grin was huge. This was good-natured teasing, to be followed, possibly, by an admonishment about setting foot as he had, literally, in the kitchen.

Sounding serious and trying to keep a straight face, he responded.

"Adapt …. Overcome …. Improvise …. Take the initiative …. Get the butter."

His tone was that of a sage drill instructor rendering advice to basic trainees on how to approach life. He was mimicking Clint Eastwood's character in the movie *Heartbreak Ridge*. After having delivered the inspirational philosophy and then pointing out that the biscuits were cooling off, he left quickly to rejoin his family, butter in hand. The kitchen staff were roaring.

"DAAAAD, you didn't just go into the kitchen, did you?" asked Susan. Problem solving dads can embarrass even 22-year-old daughters.

"I don't think there's a problem," Andrew observed. "All the people who came out of the kitchen and are looking at us are smiling."

Within minutes a huge man approached their table. He was clearly a manager. "When can you start?" he asked.

During the meal, their server and any other servers who came near the table extended big smiles and/or a pat on the back in passing, or stopped to share an entertaining comment. It was smiles all around. Involuntary ones. They all had shared something positive. *It would be great to have moments like that occur more often*, thought Mick. *Adapt and overcome. Don't let the Lenas of the world get you down.*

24 RECYCLING MATH HUMOR

It's Tuesday, the professor thought as he got up. There was a small sense of relief. It was Mondays and Thursdays that were the special days for homeowners in his neighborhood. For some reason, he began to muse about how he stood in comparison to his neighbors, in terms of the alertness level needed to be successful at the ritual practices of city living. For example, he knew that the garbage container needed to be out on the curb every Thursday morning. Further, it was a good idea to put it out on Wednesday night, in case the garbage truck came especially early in the morning. Garbage pickup was no problem for Mick. He had mastered garbage day.

The real challenge was recycling days, which occurred on Monday midmornings. But not every Monday. The recycling rule was not simple, which really means that he'd never made an effort to find out what it was. *The first and third Mondays of the month* – that, now, would have been an example of a simple rule. But the actual rule was more on the order of every other Monday. But he didn't think that that was quite it either.

He was a little embarrassed that he had never put forth much effort to learn the recycling day algorithm. Fortunately, the city put signs up on appropriate weekends to remind people that pickup would be on the coming Monday. The signs even got put up most of the time.

If he had had a really high residential IQ (RIQ), he would have either learned the algorithm or kept the pick-up schedule mailings that the city sent out. Instead, when he got up on a Monday morning, he followed his own algorithm. He asked himself if the prior Monday had been a recycling day. If the answer was yes, he knew that today was not a recycling day. If the answer was no, he'd peek out a window to see what his neighbors had done.

It was the peeking part of his ritual that made him wonder about how his RIQ score compared to the scores of his neighbors. Neighbors in some sections of the neighborhood consistently did not put recycling bins out on collection days, indicative of a low RIQ. Mick lived in a high RIQ section, where most neighbors could be counted on to have their bins out on the right days. Of course, if the only yard he saw with bins out belonged to one particular neighbor, he interpreted this as a low-reliability signal because that neighbor put his out almost every Monday. If that were the only yard with bins out, Mick would put his own bins out only if one was at least half-full.

Yesterday had been not only a Monday, but also Presidents' Day. When Mick was a kid, Washington's and Lincoln's birthdays were celebrated on different days. It was a more patriotic time, and people were aware of which holiday it was. But now, all presidents were honored on the same day. Even Carter. To many, the holiday would slip by unnoticed, in which case the only real impact of Presidents' Day was that city workers would have the day off.

Of course, any holiday observed on a recycling-day Monday would make Tuesday the recycling day. And if the holiday passed unnoticed, like Presidents' Day would to many, only people with high RIQs would realize that Tuesday was a recycling day. Mick knew that yesterday had been Presidents' Day, but didn't connect the dots to recognize that he should follow his Monday ritual on a Tuesday.

As he backed the car out the driveway, he noticed that neighbors had put their recycling bins out. He connected the dots. Mick turned off the ignition, got out, and put the newspaper bin and the bottles and cans bin out on the curb. *There is weak evidence that I have a respectable RIQ,* he quipped to himself; his algorithm derived from the fact that he was typically the last one up in his part of the neighborhood, and it worked. And he certainly had a higher RIQ than that one neighbor who lived down the street.

On the other hand, maybe he didn't live in a high RIQ section at all. Maybe his part of the neighborhood had only one person with a high RIQ, the guy who was typically the first one up in the morning. Stan. *Maybe all of his neighbors – except that one, of course – look out their windows to see what Stan did,* he thought. Such intricate goings-on would make it difficult to estimate the percentile ranking of his own RIQ score. Mathematicians have the strangest recreational thoughts.

As he drove downtown, Mick's thoughts turned from RIQ to breakfast. He could picture a waffle, and it made him smile. He soon entered Jack's and sat down at the counter next to Chemistry Department colleague Frank Kinard.

Frank was reading an issue a newspaper and eating his usual morning fare, a breakfast biscuit. After Mick ordered a waffle, sausage and coffee, Frank slid the newspaper to directly under the mathematician's nose. The chemist was pointing to an article that he wanted him to read. It contained depressing news, humor and irony, all at the same time. Some state governor was dismayed because less than one student in five had passed a statewide, grade-level exam. For those wanting to know more, the article helpfully pointed out that the exact figure happened to be 22%. Also, it was a math exam. Frank could tell from Mick's face that the arithmetic contradiction hadn't registered. "You missed it," he observed, as he pointed to two locations in the article. Mick had skimmed the article, possibly just as critically as some proofreader had. After ratcheting up his alertness level, the professor looked a second time. Frank's laughter quickly followed Mick's.

After breakfast, the math professor made haste to his office. He had to give a final exam at 8 am. Both of his sections of MATH 104 would take the final exam at the same time, but in different rooms. Mick would proctor one section and, across the hall, a colleague would proctor the other.

After the professor had distributed the exams in his room and given instructions, one of the students came forward. She had forgotten to bring her textbook. This was important because students would need to use some of the statistics tables that were in the back of the book. She was frantic.

"Don't worry" said the professor reassuringly. "I can get you a loaner book for the exam. The main math office is just down the hall. Go ahead and start the exam with the others. I'll be back in less than a minute."

As Mick went out the door to get the book, a student from his other section, a student who rarely attended, emerged from the room on the opposite side of the hall. He and Mick were face to face.

"I forgot to bring my calculator for the test in that room," the student informed the professor. "Are you in the Math Department?"

"Yes," the professor responded with a poker face.

"Could you tell me where the Math Department is? Maybe I can borrow a calculator from them."

"Sure. I'm going there now. Come with me."

25 ON THE CONSEQUENCES OF COMPULSIVE TINKERING

It was that time of the year. The College of Charleston was hosting the Math Meet once again. The event was much the same as a track meet or a swim meet, except that the competitions were in mathematics. There were relays, medleys, individual and team events; and trophies were awarded for first, second and third place. The contestants were students from high schools and middle schools from across the Southeast. There were lots of events, which meant there were lots of trophies, which meant there were lots of happy contestants. Some of he quickest young math minds in the region came to Charleston every year to compete.

The College had hosted the event for nearly thirty years. One reason for hosting was magnanimous. Organizing and hosting the meet provided good service to the math community and good nurture to the future math professionals. The other was to identify bright math minds for the purpose of recruiting. Of course, the best of the best were likely to attend the ivies, or institutions like Duke or Stanford. Still, the College did get some students who had been contestants. The math department even hired a former participant once, after he had completed his Ph.D. at Duke.

Every year, one of the event's time slots was set aside for a featured speaker. This year it would be Mick. As always, the speaker's mission was to give a talk about mathematics that would be accessible and interesting to both the contestants and their teachers. He decided to talk about how several concepts typically learned in a first course in probability or statistics had direct application to quality control at a chemical manufacturing plant. There was nothing about the math content in the presentation that would be beyond high school students. Further, someone could understand the talk without really knowing anything about chemistry. The fact that he was the one giving the talk was proof enough of that, the professor would admit freely.

One of the beauties of probability and statistics that he hit hard in the talk was their broad applicability as practical tools to other disciplines. He pointed out that statistics consultants needed to be willing to learn enough about the client activity to help address the problem at hand. And he shared with the audience that consulting brought with it two really nice intangible personal benefits. One was background broadening. The other was the satisfaction of helping a manufacturer make a better product, or better make a product, or make the same product more cheaply, or, well, just helping make the world be a little bit better than it was before.

One part of the presentation involved the quincunx, a device students always found captivating. After explaining that a chemical recipe would be followed to make a batch of polymer, he centered the funnel directly above the column numbered 5 and told the audience to imagine that when a ball bounced to the left after hitting the first pin, that that meant that chemical A had been under-dispensed, and if it bounced to the right, the chemical had been over-dispensed, and so on. After explaining that the numbers on the columns would represent some variable of interest, such as the viscosity of a completed batch, he dropped more than 40 balls down the funnel.

The outcome was predictable. The 5-column (directly under the funnel) was stacked higher with balls than any other column. The next-best populated columns were the 4-column and the 6-column, each with a similar number of balls. Then came the 3-column and the 7-column, each with a similar number of balls, and so on. The arrangement of balls in the columns made for a nice, roughly symmetric, mound shape, with the 5-column being the tallest column of the mound. The professor explained that for this polymer product, a viscosity of five meant that an ideal batch had just been made, and that five, in fact, was the company's target value for viscosity.

He went on to explain how some of the mathematical tools they were learning about in high school were used in modeling the probability with which a ball would fall into a given column. He also emphasized that viscosity, the length of an adult alligator, and, in fact, lots of variables in real-life had mound-shaped distributions, as a consequence of totaling the effects of lots of chance inputs. He talked about how normal distributions (bell curves) were what you got as a consequence of letting the number of inputs increase.

Actually, most of what he had explained up to that point had the purpose of setting up the demonstration of a practical lesson. Mick released a horizontal bar that sealed off the bottom half of the columns from the top half. Now more balls could be dropped into the funnel and they would fall into the columns to form a new distribution that wouldn't affect the mound shape of the balls already dropped.

The professor then introduced himself as Harry, one of the operators at the chemical plant. Harry, who meant well, was a compulsive tinkerer. He intended to *help* the equipment out whenever it produced an imperfect batch of polymer. With the funnel still directly above the 5-column, he dropped a ball, which ultimately fell into the 7-column. Reasoning that the equipment had just made a batch whose viscosity exceeded the target value by two units, he moved the funnel to above the 3-column, and dropped another ball. When the ball landed in the 2-column, Harry decided that he now had to adjust for the equipment producing a batch with viscosity three units too low, and he moved the funnel from atop the 3-column to atop the 6-column. The professor continued making batches and making *corrections* until the shape of the distribution that was developing as a result of Harry's compulsive tinkering was clear to the eye and clear in contrast with the first distribution.

The professor asked if anyone could say what was different about the two distributions. A student who got it spoke up, and the professor reinforced the observation. With the equipment left alone, there was a heavy concentration of viscosities near the target value of five, but with Harry's constant re-centering of the equipment, there was less of a concentration near five and, consequently, a higher percentage of viscosities far out in the left and right tails.

Explaining that the vertical line that separated the 2-column from the 3-column was the line of demarcation that separated defective batches from acceptable batches, he observed that even if the equipment were centered slightly off-target, say at the 4-column, fewer defective batches would result than when Harry was left in charge. In statistical process control, Harry's method of controlling the process by reacting to individual values was called *overcontrol*. The next part of the presentation was devoted to showing them one of the simpler kinds of control charts and how it used data to decide appropriate times to signal warnings that the process might need to be recentered.

He concluded by hyping the moral of the story, which was that statistics provided tools that would help people make changes at the right times, whereas making process changes in reaction to individual events or to isolated incidents would increase variation in the output and, even worse, increase the number of undesirable outcomes. In life it was all too common for people to use examples of isolated, undesirable outcomes to argue for changing entire systems. Of course, in real life, one could always find individual examples of just about anything, which always made it possible for well-meaning people, or for hucksters, to argue for making global changes. The professor closed with the message that he wanted students to know the consequences of overcontrol, to recognize overcontrol when they saw it, and to know better than to do it.

During the question and answer period that followed the talk, the most interesting comment to come from the audience was from a coach of one the out-of-state teams, who observed that the first thing that had popped into his head as an example of the negative consequences of overcontrol was overparenting. Mick's own personal favorite egregious example of a compulsive tinkerer was the U. S. Congress.

26 WE RANDOMLY PICKED YOU

He had brought nothing to read. The professor was at a small airport, waiting at a boarding gate, killing time. Here was a chance to chew on one of those fond old memories. Actually, the one he settled on was fairly recent, an incident that began at a card game.

He and Libby had been playing cards with a group of friends. For years, but not in the summers, the group had gotten together on a once-a-month basis. The location and host rotated through the group, and hosts always provided red and white wine, soft drinks, and snacks. The card game was Euchre, and while it was always more fun to win than lose, it didn't matter much who won. The chatter during games was as enjoyed as the game itself. The particular evening he was remembering ended, as usual, with dessert and decaf coffee. The choices that evening were ice cream and strawberries, or cheesecake, or both of course. As usual, all was delicious.

At one of the card tables, Mick and two other men were talking and eating. One was a WWII veteran and the other had been a child in Germany during that war. Taking advantage of an opportunity to get a rare glimpse into history, Mick asked the latter what he remembered about growing up in Germany during the war. From then on, it was all guy talk at that table.

At some point, the topic had shifted to what they perceived to be an undeniable, widespread deterioration in customer service. Mick's contribution was describing a bad customer service experience he'd had when trying to obtain hearing aids through the Veterans' Administration. Listening to the story aroused the interest of the WWII vet. Then, based on his own knowledge and experiences in dealing with the VA, the older vet began to offer advice and counsel to a fellow former soldier. At some point, the man who'd grown up in Germany asked what Mick had done in Vietnam. After the latter shared some of that personal history with them, the WWII vet decided to open up a little about his experiences as a soldier. What he shared was a wonderful glimpse into history – and a compelling personal story. Everyone there that night knew that their friend had been a POW in WWII. But he had never talked about it before.

Several days later he told Mick that he'd kept a written record of his war experiences and that only immediate family had ever been allowed to read it. He asked if Mick would like to borrow it. The guidelines were simple. Mick could read it, but was not to let anyone else do so. And so, the professor was lucky enough to read the chronicle of a second WWII vet, a friend of his who shall go unnamed. Fate's roll of the dice had changed his life forever. The friend would be embarrassed for his name to be known.

He was one of the many young, brave souls who parachuted into Europe for the D-Day invasion. He drew one of the short straws that life sometimes dishes out. He landed in a pond that was in the vicinity of German soldiers and was captured shortly after landing. So while comrades of his were fighting and dying and bringing the war to an end, he was marking time in a POW camp in Poland, a camp where he says prisoners were not treated harshly.

When the end of the war was imminent, German soldiers began marching the prisoners toward Germany. One evening the march stopped at a farm, where prisoners slept the night under guard. He managed to hide deeply enough inside a haystack to be beyond bayonet reach. Having grown up on a farm, he knew his way around haystacks, and how to be deep but also safe. He made his way to freedom after the march resumed the next day. As with what was said earlier about *Over Here* (Don Johnston's chronicle), the summary just given of the former POW's experiences doesn't do justice to that chronicle.

Again, the friend would have been embarrassed to be named. It was obvious why. Guilt. The undeserved guilt that is natural for soldiers to feel. People his friend had trained with were fighting and bringing the war to an end – and dying. And he wasn't. Did someone die instead of him? Did someone die because he was not there to keep it from happening? Had anyone been maimed who wouldn't have otherwise? They were fighting and dying, and he wasn't helping.

The friend didn't need to spell it out. That he was dealing with the *Why me* question was easy enough to recognize. The guilt feelings had to have been severe. All that time spent in the POW camp. All those soldiers who died while he was there. *Random events with lifetime consequences*, thought Mick. He was glad that his friend had opened up after the card game. Maybe talking about it had eased the burden a little.

Several days after having read and returned the chronicle, Mick happened to bump into his friend. The professor took the occasion to comment that he strongly suspected that the reluctance to talk about WWII was from guilt.

"You could tell that?" the friend asked, a bit amazed. The friend went on to say that that was the case; further that the guilt was very strong.

"You know, it's not your fault that you were captured," the younger vet observed. The former POW responded that he understood that at the rational level, but that the guilt was still there nonetheless.

A loud voice transported Mick from somber thoughts back to the present and the fact that he was at a boarding gate. Boarding was to begin. The numbers of the first rows to board were announced.

So that terrorists will never know, the name of the airport is omitted. Suffice it to say that the airport was not in a major city and that the planes that took off and landed there were not exactly jumbo jets.

The professor was now smiling, glad that he was lucky enough to know the former POW. The latter was duty-honor-country in demeanor, and a real gentleman. The boarding pass was in Mick's hand and he was waiting for his row to be announced.

Two uniformed airport police officers – a man and a woman – approached the professor.

"Excuse me, sir," said the man. "Please come with us. You have been selected for an extra security check."

As if someone had just told him a clever joke, Mick began smiling. Broadly. Thinking that he might be viewed as a threat made him start laughing. He couldn't help himself. A feeling formed – that he was someone brought up from the audience to participate in a comedy skit.

"You picked *me?*" The tone of the question was not judgmental. Rather, it reflected incredulous amusement.

"Sir," the man continued, "this is serious. You have been randomly selected for extra screening. You'll have to come with us. Don't worry. You'll be back in time to board."

"OK, sure," the professor said placatingly. He had stopped laughing and was trying to display the level of seriousness warranted by the occasion. Besides, there was no choice but to participate in the exercise.

They led him to a counter within sight of the boarding gate and had him put his carry-on bag on the counter. The woman looked through the bag while the man wanded him.

As the exercise was wrapping up, Mick decided that in order for him to rate the show, the audience needed to interact with the cast members.

"I have a question for you. You said that I was *randomly* selected. I'm a mathematician and am curious about how you use the word *random*. Is there some combination of letters or digits on my boarding pass that caused you to pick me?"

"No, sir," said the man. "WE (as in *the two of them*) randomly selected YOU. We always select a random man and a random woman from each departing flight. And between you and me, we always pick someone it won't take long to check."

Wow! From a mathematical point of view, picking two passengers using a truly random process – as in drawing straws – would be a terribly unproductive strategy. That is, if the goal was to really catch bad guys. With such a strategy, the bad guys would not feel pressed to try to masquerade as happy, little old grandmothers from Des Moines. Moreover, if the bad guys discovered that the strategy was to pick two people who were easy to check, the risk would be higher than when using the pick-two-at-random strategy.

Of course there was the possibility that, on those occasions when they really were suspicious about someone in particular, the gendarmes would use common sense about whom to select for their random screening. Mick hoped that picking two people who were easy to check was what they did when they were not suspicious about anyone in particular. He hoped so, anyway.

27 LUNCH WITH COLLEAGUES IS GOOD THERAPY

One of Mick's favorite pleasures was lunch with colleagues. There was a lunch group in the math department. The faces in the group would change over the years due to retirements, and even semester-by-semester, depending on individual teaching schedules. What was constant was light-hearted discussion, often about department business or mathematics, but not necessarily. Faculty members would talk about their kids, about hassles getting a driver's license renewed, about current events, and about many things. The most serious problems of the world were solved during lunch. For example, one day the topic of discussion might be the question, *Should every undergraduate be required to take a statistics course?* Mick could be counted on to make a case for an answer of *yes*. Naturally some would agree, some would disagree, and another consequential issue of world proportions would be thoroughly explored. The same players would take the same positions when, over lunch, the same question would crop up again in a year or two. Recurrent questions would always be debated afresh. Whether anyone recognized that the discussion was a rehash could not be discerned. The passion level was always the same.

Another common lunchtime theme was grousing, which was occurring as Mick bit into his sandwich one day.

"Man, these tenure panel meetings are getting to me."

"Some people really like to hear themselves talk. Yesterday, I'm sure I was not the only one thinking *get on with it, already*. Of course, we all sit there and listen because we're all too polite to say anything."

"I agree. What wears me out the most is fine-tuning the rough draft of the *panel letter* that's supposed to summarize all of the panel's discussion. Everywhere there's an adjective, we seem to spend hours deciding which adjective is the best one to use."

"Yeah. Should we say this guy's teaching is *superb*, or that it's *excellent?* Some people are willing to debate minutia like that forever."

"Actually, it was long ago that we went well beyond the point of diminishing returns in arguing about stuff like that. And every year it gets worse. The crazy part is that when the panel letter gets read by the College's Tenure and Promotion Panel, whichever of the adjectives were left after the others were voted off the island won't make a difference. The College T and P Panel has no idea how much debate went on over pointless nuances. It's like we've taken a numerical reading in a lab experiment where we're supposed to report what number we got. We spend hours debating what the fifth decimal place is, and the people who read the report only care about the first two places. Much ado about nothing."

"Well, it's not just that the people in the department bring it on themselves. The administration has also contributed at lot by how it keeps changing the rules."

"That's true. Remember when it came down from on high that to get tenure, somebody had to be categorized as *exemplary* at something?"

"You're right. So one person thinks the candidate's teaching is exemplary. And somebody else says, no, it's his research that's exemplary. And somebody else thinks his teaching, research and service are all exemplary. And not only do we have to take time to hash that out, we have to summarize for the panel letter the discussion of what we have decided is exemplary. So it can come out like a box score: three people think teaching is exemplary, five think ..., and so on."

"I don't know how much longer I can stand these meetings."

"Well, how many years until you retire?"

"I could do it in twenty. But I'll never retire. I can't afford to. More than likely I will die in the classroom, probably writing on the blackboard, probably teaching calculus."

"Well, just make sure the last thing you write is correct. If you're solving an indefinite integral problem, make sure not to die before you write the "$+C$" that goes at the end. Oh, don't worry. Sure, we'd still all come to your funeral. But we'll also take off a point."

"On a different note, I've got a question for the people with kids who are older than mine. My two are at an age where they've started arguing and fighting a lot. Actually fighting. We stop them from hitting each other, of course. My wife and I do not believe in violence and are not quite sure how to handle this. Should we let them fight? To those of you with older kids, what should we do?"

"My brother and I fought all the time when we were growing up and it didn't hurt us. We had some really good wrestling matches. After a while, though, when he started winning, I played the *girl* card and said we'd have to stop. Those wrestling matches actually gave us some great memories. We laugh about it now."

"I agree. Let them tussle within reason. There is a how-to-get-along aspect to it. They'll learn social skills that will help them when they grow up. Maybe it's a little like what bear and lion cubs get out of play-fighting."

"You mean like establishing a chain of dominance within a pride? You were an only child. What do you know about getting along with siblings?"

"Well, besides our having raised two kids, I get a lot of pointers from watching Animal Planet."

28 WHEELING FOR DOLLARS

It had been decades since Mick had contacted lawyers in the Charleston area to let them know he was around if they need help with statistics. Ever since, cases had found their way to him. The College was comfortable in allowing him to consult. The main understanding was that it should never interfere with his day job. And it never did.

Maybe he'd get a case every other year. The professor viewed consulting as a hobby. But unlike stamp collecting, it was a win-win for multiple stakeholders. His students benefited by getting interesting classroom examples. The College benefited by getting good PR. Decision makers from a variety of settings had been helped by his mathematical and statistical analyses. And yes, he made some money on the side. But more than the monetary benefit, the hobby enriched his professional depth, not just in regard to teaching. The Hettmansperger-Norton statistical test for patterned alternatives might never have come to exist if it hadn't been for the first age discrimination suit Mick had ever worked. There were presentations at meetings and journal articles that wouldn't have existed otherwise. And there were people helped, like that young woman, the med student who was accused of cheating on an important end-of-year exam. Her future career was likely saved by his testimony – about how to interpret some probabilities – before an honor council. In the end, they had found her not guilty, which was the right call in light of the statistical case.

The hobby had not been without annoyances. There had been a good number of deans, provosts and presidents over the years, and new administrations, that made new paper-trail rules about faculty members doing outside consulting. New administrations liked to reinvent the wheel, uninterested in whether there were any old rules or understandings. Or if someone, somewhere abused the existing rules, new rules that changed the system for everyone might be created. Physicists should stop to consider whether wheel reinvention in higher education meets the definition of a perpetual motion machine.

One day, while keeping an eye on web sites that showed the kinds of jobs in statistics that were available to his students, he came across an ad for a consulting statistician. The ad gave not the slightest hint of the area in which statistics would be applied. But, on a lark, and thinking that an interesting gig might come out of it, he submitted his resume. Shortly thereafter he received a call asking if he wanted to be a court appointed expert witness.

He had testified in court before, and had given depositions, but this was the first time he would be a court appointed expert. It also would be the first time he would see how Medicare might interact disapprovingly with a healthcare provider. An audit had been conducted of a company that provided motorized wheelchairs to qualified individuals. Medicare had paid the company for 188 claims over a period of years. Give or take a few bucks for wheelchair add-ons, each reimbursement was for close to $4200. Altogether, that's ... , well, ... the multiplication is left to the reader. Medicare could have examined the validity of all 188 claims, which would have been time consuming. Instead, a simple random sample of 50 claims were carefully examined. It was found that 46 of the 50 should not have been paid. Medicare wanted the company to pay it back for what it called the *overpayment*. The case was being heard by an administrative law judge. Separate from the case, fraud charges were being pursued.

Mick's role was to give his opinion as to whether the statistical method that Medicare used to estimate the total overpayment that accrued from all claims was consistent with sound statistical practice. For this, there would be verbal testimony and a written report. The healthcare provider also had hired its own expert, Dr. Hanover Bunque, Ph.D., who would likewise critique Medicare's statistical methods. Finally, the two experts could critique each other's critique, a sort of High Noon for dueling experts. Bunque, um, had a history of working for defendant companies in such cases.

The salient facts were that 50 claims had been sampled from a population of 188 claims, that the software used to determine which 50 would be in the sample could be trusted to obtain random samples, and that 92% of the claims (46 out of 50) shouldn't have been paid. *Facts* may be too strong a word. For example, it was not Mick's job to determine whether 46 was the correct number. Investigating which claims were valid and which ones weren't was not his area of expertise.

The professor did not need to go to Florida to testify in person. It was set up as a telephone conference. Parties who were to testify on a given day were told to call in at an appointed time and wait until called upon to testify. This meant that on the first day, Mick spent over two hours in his office listening on his cell phone to the testimony of others. Some of the testimony he heard didn't affect his part of the case, which meant that his mind could wander. *Holy cow, what will my cell phone bill be this month? Was there a potential lawsuit in the future regarding the link between brain cancer and prolonged cell phone use? Will Medicare still exist when I need it? Will the St. Louis Cardinals go cold now that they have clinched their division?*

When his turn finally came, he had to be sworn in. "Raise your right hand," the lady said. *How would they know?* he thought. "Is your hand raised?" she asked. *Aha, they will know, provided I tell them the truth.* The reader should not fear. Truth be told, his hand was raised for the swearing in. There were several reasons why testifying over the phone sure beat being in court. But of course, there is no need to talk about all of the multitasking he did while he was on the phone. Ah well, back to the case.

In extrapolating from the sample to the population, Medicare had arrived at a conservative estimate of 162 (86.2% of 188) for the number of claims in the population that shouldn't have been paid. This conservative estimate led to a dollar amount of more than $674,000 that Medicare wanted back.

There were several statistical issues raised by the healthcare provider's expert. One had to do with using a sample size of 50. A Medicare document actually explained the choice. The rule that was used to decide the sample size was one that 90% of the time would make the proportion of overpaid claims in the sample be within .10 of the proportion of overpaid claims in the population. It had been easy enough to check. Actually, Mick found that for a population of 188 claims, using a sample size of 50 would achieve the goal at least 93% of the time, not just 90% of the time. That meant that a sample size of 50 might have been slightly larger than necessary to meet the goal. Nevertheless, Bunque still was concerned. His report cited a well-known sample size formula in a well-known book. Not only was the formula not for the setting at hand, the report maintained that under certain assumptions (which would have changed the goal of having the sample proportion be within .10 of the population proportion 90% of the time), the formula would lead to a sample size of 352. *Where did they get this guy?*

When the professor responded that a sample size clearly couldn't exceed the population size, Bunque, um, explained that this item in his report was being misinterpreted, that the number 352 provided simply an illustration of the application of the formula, and not the sample size that should be applied to the case at hand. Mick then listened as the other expert challenged that 50 might have been chosen because it was a nice round number.

"Further," he posed, "what if the formula (that same inappropriate formula) had said the sample size ought to be 45? Then 50 would be too big a sample size."

Where did they get this guy? Mick responded by making the observation that there was never a statistical downside to using a sample size that was larger than it needed to be.

On Mick's second day of testimony, the judge asked the professor to explain some numbers that appeared on a Medicare Statistical Analysis Department document. The document identified numbers that had been used to arrive at the amount of just over $674,000 that Medicare wanted back from the company. Actually, during Mick's testimony on the preceding day, the judge had told him that this would be coming, so Mick had worked on an explanation as if he were developing a lesson plan. The professor thought that Medicare's statistical method of arriving at a conservative estimate of the true overpayment was easy enough to understand if he chose his words carefully. It was part of the expert's job to choose words that would explain the big picture without going pedantic on statistics, just like it was also part of the job to avoid using words like *pedantic*.

"First of all," asked the judge, "do you see the number 162 on that document? It is labeled as a '90% lower bound for the number of overpayments.' Could you explain what that is?"

"This is Medicare's conservative estimate for the number of claims in the population that shouldn't have been paid. I checked to see if their terminology meant what I assumed it did. As a result, I figured out where 162 comes from, and I know what they mean when they say '90%.' However, their terminology is nonstandard. Most people in statistics would call 162 a lower 95% confidence bound for the number of overpayments. To explain what that means, first I'd like to talk about how 162 is conservative, and then try to clarify just how conservative a lower 95% confidence bound is.

"One straightforward way they could estimate the number of overpayments in the population is to use the fact that 92% of the claims in the sample were overpayments. That figure comes from dividing 46 by 50. They could then extrapolate to the population by estimating that about 92% of 188 claims in the population would be overpayments."

"Hold on ... . That would be about 173?" queried the judge. The professor double-checked the computation.

"Yes. But by extrapolating that way, there is a problem. There is a good chance that the percentage of overpayments in the sample would be higher than the actual population percentage. If that happens, they would wind up asking for too much money back. So they're going to back off from 173 and use something smaller.

"Now, the method that was used to come up with 162 started with the fact that 46 out of 50 claims in the sample shouldn't have been paid. If instead there had been 44, or 45, or 47 out of the 50 that shouldn't have been paid, then a number slightly different than 162 would have resulted. But the statistical rule they used that begat 162 from 46 is a rule that will produce an underestimate of the actual number of overpayments 95% of the time. About five percent of the time, there will be a fluky sample that leads to an overestimate of the number of overpayments. Nobody knows if this sample is one of the 5% or one of the 95%. However, Medicare does know that by using this method, 95% of the time when they take their sample, their estimate of the number of overpayments in the population will be lower than the actual number of overpayments."

"OK," said the judge. Then the document shows that to get their total estimated overpayment of $674,046.03 they have added $192,970.99 and $481,075.04. The smaller number is labeled the 'sum of sample overpayments'. Next to the figure $481,075.04, they have a formula. Do you see where that is? Could you shed some light on what's going on there?"

"Yes, your honor. First they came up with the low-ball estimate that 162 claims should not have been paid. They know there were 46 such claims in the sample, and that those 46 claims totaled $192,970.99. That much money they know for sure that they want back. That's the first of the two numbers cited. But they want payment back for 162 claims altogether. That leaves 162 minus 46, or 116 other such claims not yet taken into account. Rather than trying to estimate a dollar figure for 116 claims, they simply found the smallest 116 claims from the un-sampled part of the population and added them up. That's where the $481,075.04 comes from."

"But wouldn't the sum of the smallest 116 claims be smaller than what they would get from a good estimate?"

"Assuredly. But they do it that way for two reasons. First, it's simple – they don't have to estimate anything. Second, it doesn't cost them much. Even with different add-ons that can go on wheelchairs for different patients, there is not much variation in wheelchair price. If there were more variation in the price of wheelchairs, they would have used a different method to come up with a dollar figure for 116 wheelchairs.

"Basically, everything they have done takes a conservative posture. The method they used to arrive at the estimated total overpayment was designed to produce an underestimate of the actual overpayment at least 95% of the time. The method they are using is understandable and sound on statistical grounds."

The judge invited the professor to listen to testimony that came after his, if he chose to. Mick listened for a few minutes, but it became clear that the statistical part of the proceedings was over. So the professor closed his cell phone. And yes, that month's phone bill did turn out to be pretty hefty. The pay was good, though.

He never learned the outcome of the case, although he thought he had a good feel for what would happen. The judge's tone had conveyed that he understood the ideas behind Medicare's statistical method, why it was conservative, and the extent to which it was conservative. Also, not all of Bunque's criticisms on statistical issues have been described. It was transparently obvious that Bunque's goal was to muddy the statistical waters by questioning anything that anyone else did, even if that anything else didn't hurt his case, and even if the point he made was silly. Criticisms raised by Bunque and by the healthcare provider's attorney about statistical issues addressed by the professor were all ignored or denied. Actually, Mick felt sad. Bunque, um, had not exactly enhanced the image of the statistics expert witnesses.

29 THAT'S MATHEMATICS

The semester had been over for a few days. His students in the math modeling class had all worked hard and the grades were pretty good. There were twelve students, mostly math majors, and he was pleased with how the class had gone. Not only had the students learned a good number of topics in discrete math, they also had had lots of practice doing proofs. Students were to try to find proofs for all of the problems that were assigned for homework. Most importantly though, a student was assigned to each problem. The student was to write his or her proof for that problem on the board the next day. Mick would critique every proof, tweak it as needed, and find flaws if there were any. This way, every student saw not only every other student's proofs, but also the common mistakes that befall beginners. The only proofs that were graded were on exams. Since what they did at the board was ungraded, students quickly became comfortable at the board and had come to view the critique regimen as practice, which was Mick's goal.

He also felt good about his introductory statistics classes. Of course, there would be some students who didn't like their grades, but that was normal. The professor looked himself up on Ratemyprofessors.com to see if there were any current extreme reactions of either kind. The ratio of frowns to smiley faces was about 3 to 1. Only one entry was new. It was a smiley face with this verbiage:

> He is a great teacher. All of the other ratings on here are people who wanted to get away with out doing anything in his class. He is very helpful and his tests are like 10 questions long.

Take him because he is easy, and he looks like the math teacher from clueless. He is a cute old man.

A cute old man, he pondered. The *cute* part was OK. But that *old* part. It was the second time that day that he had been prompted to think about aging. The first time was when he had put on his hearing aids in the morning. He'd had these particular aids for about a year, courtesy of the VA. One thing that made this particular aid interesting was that when a battery was inserted and the battery door closed, the aid would play four notes after a ten-second delay. Da da da da. The ten-second delay allowed the wearer enough time to put the aid in the ear, hear the four notes, and know that the battery was good and that the aid was working.

That morning, it had occurred to Mick that, for about a year, he'd heard those notes on a daily basis, once in each year, and had never thought about them before. Until that day. Why four notes? Da da da da. Why not just a single tone? Why would a hearing aid company choose those four notes? After a second or two of thought, he believed he had the answer.

> When I get older losing my hair,
> Many years from now,[xi]

At least I'm not 64 yet, he thought.

The rest of the morning and afternoon, Mick worked on a problem in mathematical statistics. It was a conjecture that had been proposed to him by colleague Gary Harrison, one of whose interests was biological modeling. For several years, Gary had been working on developing probability models for the number of patients who would be in a hospital on any given day. There was an obvious application to hospital staffing. Gary had a conjecture. Specifically, he suspected that if one made reasonable mathematical assumptions about the random daily arrival of patients, about the probabilities involved in how patients moved from one patient class to another – e.g., from ICU into long-term care – and about the probabilities involved in a patient's going home from a given patient class, then the total number of patients in the hospital on a random day would have a Poisson distribution, a distribution well known to statisticians. By the end of the afternoon, Mick had proved that Gary's conjecture was true, provided an assumption was added. The assumption, called a *steady-state assumption*, also was necessary. In everyday English that almost describes the concept, the assumption was that the hospital needed to have been in operation for a long time. In practical (and mathematically correct) terms, the consequence of adding the assumption was that the number of patients on any given day would have an approximate Poisson distribution, once the hospital had been in operation beyond a phase-in period.

Mick had just gone over the proof with Gary, and both were smiling. The older Mick got, the more he wanted his mathematics to make a difference. Of course, there was nothing wrong with, for its own sake, developing beautiful new mathematics that had no obvious practical application. Some of his work over the years had been along those lines. In mathematics, some insights, and even proofs, had a beauty or genius to them. One could be awed by a theorem or proof, just as one could be awed by a particularly beautiful painting – art without apparent application other than to please. Or to give other artists ideas. In proving Gary's conjecture, he had used statistical theory to validate something having a practical application – determining the number of staff needed to safely handle the daily workload. Not only did he feel good about recognizing that a steady state assumption was needed, the proof was cute. And sweetening the pot even more was that one of his pet interests was in applying statistics to decision-making that improves customer service. Doing mathematics just didn't get any better.

The professor drove home in excellent spirits, still enjoying the mathematical high. *Math highs are very high and the lows are very low,* he thought. He was reminded about a night some months earlier when he and math colleague Garrett Mitchener had been driving to dinner. Garrett, a relative newcomer to the math department, was feeling low. He had been working on a research problem and the computer wasn't producing the kind of simulation output he was expecting. Garrett was confident that his math was right, but he couldn't find the glitch that was keeping the computer from producing the expected results.

"It's really disconcerting," admitted Garrett. His frustration was impossible to miss.

"That's mathematics," Mick had observed in a mentor-like tone. For whatever reason, the Broadway song *That's Entertainment* popped into his head. He sang to Garrett, making up lyrics as he went.

Proving theorems,
That's what we do.

When we can't,
Then we're feelin' blue.
When we can,
Then we're smilin' at you.
That's mathematics.

Mick sang the words again as he drove home on the James Island Connector.

He arrived home before Libby. After sorting through the mail, he marked time by trying to get on the internet. No dice. The usual fix – unplugging the modem for a while and rebooting – didn't cure the problem. He called the internet provider and got the expected recording: *For service related issues, press 1, for billing, press 2,* and so on.

In hopes of getting a human quickly, he didn't press anything. After a while, a voice said that a customer service person would be with him shortly, and that for quality assurance purposes, his call may be monitored. He found himself listening, not to mood music, which would have been bad enough, but to advertisements and promotions. After about 40 seconds, the promos stopped and he heard: *Your call is important to us. All of our customer service representatives are busy right now. Your call will be answered in the order in which it was received.* Then came another 40 seconds of the same promos, followed by the same three-sentence message.

Sometimes he would make sarcastic remarks to telephone customer service recordings when he was on hold. It was self-entertainment that actually made him feel better. A recording couldn't take offense, so what was the harm.

After hearing the message for the fourth or fifth time, he decided he'd had enough. The high he had been on before encountering computer problems made him want to unload creatively before hanging up. His tone was stern.

"Yeah right, my call is important to you. I've heard this message five times now. If you tell me one more time that my call is important, I'm going to cancel my service. That's not what I called up for, but it's just what I'll do. And if you don't believe me, … try me."

He took pleasure in being silly with the false dare. Part of the pleasure was that John Wayne couldn't have delivered the words in a more ominous tone.

Immediately after the dare, absolutely immediately, a customer service person came on the line. Mick described the problem to her and she said she'd connect him to someone who would fix the problem. Mick never did get to talk to a fixer. He was put on hold briefly, during which time he could see that the fixer had fixed the problem. Mick waited for about a minute to relate to someone that everything now worked, but when no one spoke up on the other end, he hung up. When Libby arrived home, he told her about the bizarre coincidence – that someone had just happened to take his call as soon as he'd delivered a dare. It made for a nice story over dinner.

The following Sunday, he and his wife were watching an episode of Monk, a weekly television series. The main character, whose last name was Monk, was a former police officer with a gift for piecing clues together. In almost every episode, police would seek his advice to help them solve crimes.

In this particular episode, instead of asking Monk to act in an advisory role on a case, they let someone else be their consultant. The competing consultant had come forward, offered to assist with the case, listened to limited information, and showed Monk up by displaying an uncanny sixth sense for how to interpret clues correctly. Monk knew that the other guy had to be "cheating" somehow. By episode's end, he had figured out how. After the bad guys had committed the crime, they called from a hotel room to order airline tickets and were put on hold. While on hold, they discussed what they had done. Monk discovered that when calls were monitored for quality assurance purposes, someone at the vendor end really could be listening during the mood music. In this case, the customer service person was the competing consultant's mother, who provided her son with all the details he needed in order to look like a detective prodigy.

The professor wondered if maybe the miracle of a coincidence with the internet provider wasn't such a miracle after all. An opportunity to find out presented itself the following a few days later when Mick had another occasion to call a service provider. After hearing the your-call-may-be-monitored announcement, and listening to the same really annoying piece of music ad nauseam, he spoke.

"This choice of music doesn't do a thing for me, and I've heard it now six times. If you expect to keep customers, you shouldn't do this to them. I'm going to count to ten." "One. Two. Four. Six. Eight."

"Hi. This is Susan. How may I help you?" interrupted the count.

30 OBJECTIVE EYES

It is natural for people whose jobs require them to do the same thing over and over, and for the service-providers or manufacturing companies that provide those jobs, to be so ingrained in the process that they will argue that nothing can be improved. *That's the way we've always done it and there is no better way.* It is a difficult attitude to overcome.

The fact that people can be tenacious when defending traditional routines and the fact that sometimes, data have to be fortified with people skills and persistence in order to persuade someone to change a routine are bullet points the professor would include if he were to make a Power Point presentation on important observations made during that sabbatical year. It was one of the three sabbaticals he took during his career at the College of Charleston, also the one that was the most fun.

As clarification for the reader, the purpose of a sabbatical is professional re-energizing. A faculty member is eligible for a sabbatical every seventh year. Of course, when God needed a break, He took the seventh day completely off, whereas a faculty member wouldn't have a sabbatical approved without first submitting a good project proposal for how to spend either one semester or a full year doing something other than the same old-same old. All of Mick's sabbaticals were spent doing research. All were fun, all had provided rejuvenation, and all had resulted in publications that were good for his career and good for College PR. Ah, but that one sabbatical was special.

Prior to submitting a proposal, he called Gregg Adams, who worked at a manufacturing plant. Gregg was a mid-level manager with a role in interpreting data for product quality assessment. He also had recently taken

Mick's graduate course in statistical process control. Moreover, the two had known one another from many years before, when Gregg was an undergraduate and taking calculus from Mick. The professor asked Gregg if his company could use him for a year.

And so it came to be that Mick would spend the year acting as a pair of objective eyes. He was to spend the first month observing and learning the manufacturing process, meet people, see what they did and how they did it, and use his knowledge of statistics and statistical process control to look for ways to improve the quality of the process and of the product. The product shall be referred to as "thread."

Interestingly, during that first month when he was learning the ropes, his input was sought once – on a mathematics question rather than a statistical question. The professor had already learned enough background information to understand the vocabulary of the problem, namely that denier (pronounced *DEN yer*, a term meaning *thread weight*) was one of the key variables the company measured. The denier of the thread on a spool was the weight, in grams, of 10,000 meters of thread. So for example, 10,000 meters of 40-denier thread weighed 40 grams. The other relevant background information was the geometry of the industrial-size 40-denier spool of thread. The thread was wrapped around a core similar to the cardboard core of a roll of paper towels. The thread, viewed as a solid with the core removed, is pictured on the next page.

The problem posed went something like this. Suppose the outside diameter of the thread on a standard 40-denier spool is 18 cm, the inside diameter of the thread is 6 cm, and the height of the thread is H (it is not necessary to know the value of H). Suppose the standard 40-denier spool contains 42,000 meters of thread. A customer who wants to buy a large shipment of 40-denier spools has warping equipment that will accept the standard core, but the outside diameter has to be 16 cm. How many meters of thread would be on such a spool?

Of course, the reader should understand that the standard spool would not contain exactly 42,000 meters of thread. This is because even the best of processes has variation that keeps things from being perfect. Thread thickness along any given length of thread would vary a little from point to point, so that thread weight would not be perfectly uniform. Further, the method of determining the length of thread on a spool involved weighing the thread on that spool. Also, the outside diameter of any given standard spool wouldn't be *exactly* 18 cm – that is, 18 followed by a decimal point followed by infinitely many repeating zeros thereafter. And finally, measuring each of these quantities introduced measurement error.

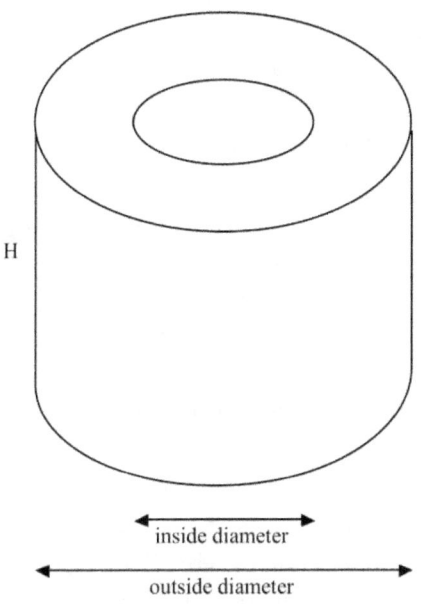

inside diameter

outside diameter

Mick worked on the problem as posed for the ideal spool, coming up with an answer of 32,083 (and one-third) meters of thread. He also provided a formula that would give the answer in terms of an arbitrary outside diameter D. He announced that there would be about 32,100 meters of thread on the kind of spool they were to make for the customer. After starting production, a routine sample of spools was taken to the testing lab where various thread characteristics were measured. Using one of the spools to confirm thread length, lab techs arrived at a figure of 32,023 meters. Gregg told the professor that upper management had concluded that the formula had to be correct and decided that Mick had to be an amazing fellow because the estimate was "off" by a very small amount, roughly 100 meters. He smiled when Gregg, who had seen Mick's computations, said he had eagerly pointed out to the upper level managers that the estimate was really only 60 meters off. The professor smiled. To him, being "off" by 60 meters out of more than 32,000, even though the number applied to just that one spool, was a sign that the company was doing a good job of controlling thread weight and of obtaining measurements in the lab. Of course, he also knew better than to make too much of what had happened on a single spool. However, as he learned more about the plant over time, he found that the single number really was telling the true story.

There couldn't have been any job more pleasant than one without a daily requirement of grunge work. He was to meet people, see what they did, see how they collected and dealt with data, and recommend changes when he saw something he felt could be improved. The only requirement was that he had to keep Gregg informed about what he was doing and thinking. Gregg, who made arrangements for Mick to meet people, also had the role of sounding board for Mick's ideas. Particularly early on, this meant providing guidance about process realities when Mick would ask naïve questions borne out of lack of familiarity with the product and its manufacturing process. Of course, asking naïve questions is inherent in any learning process. Mick spent much of the first month awestruck by the quality and professionalism of the people he met. It quickly became obvious that the company was very careful about whom it hired.

One of the professor's major contributions came after spending some time watching the goings-on in the test lab. One of the tests they performed, a tension test, particularly piqued his interest. The amount of tension needed on the loose thread end to make the thread pull freely from the spool without snagging was an important spool characteristic to the company and its customers. It will be helpful to point out that spools manufactured by the company did not dispense thread in the same way a small spool meant for a household sewing machine would. The wooden or plastic core of the latter would spin on the spool pin as the sewing machine pulled the thread in a direction perpendicular to the hole in the core, whereas the core of the former was immobile and thread was pulled from the spool in a direction parallel to the hole in the core. In making fabric, a machine could pull thread from hundreds of spools at the same time and, in the spirit of a dance team, all spools needed to react to a given amount of tension pull in the same way. Just as a dancer who is out of step would stand out, a thread line that acted differently would have visible consequences in the fabric.

Mick observed that when a routine sample of spools was taken to the testing lab, a spool used for a tension test was immobilized on a vertical post, one that protruded downward from a frame in the fashion of a perfectly cylindrical stalagmite. A lab tech would pull the loose thread end down until the end was just below the bottom of the spool, clip a small weight to the loose thread end and let go. As many times as necessary, the tech would repeat, using successively heavier weights, until the thread would fall from the spool without snagging – meaning that the thread wouldn't stop falling until the weight hit the floor. In recognition that every process had variation, a reading had to be confirmed before it would become recorded as the value obtained from the tension test. With scissors, the tech would cut off the thread that had fallen, pull the new loose end down to just below the bottom of the spool, and repeat the process, again attaching a weight that was too small. The process would continue until the lab tech got the same value at least twice in three replications.

Mick noticed that the tech showed no interest in where the loose end contacted the spool. The loose end could have hung from the top of the spool, from where the thread wound downward, hung from the bottom of the spool, from where the thread wound upward, or hung from anywhere between the top and bottom. The first two scenarios are pictured below.

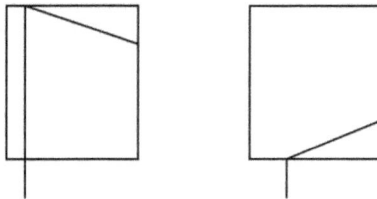

It seemed to Mick that, on average, it ought to take more weight to bring down thread hanging from the bottom because there would be drag to overcome, whereas thread hanging from the top was almost primed to fall because, with the thread winding downward, there would be very little drag.

He brought his observations to Gregg and proposed doing a test that involved obtaining four tension measurements from each of four hanging-thread scenarios. These scenarios are pictured on the next page. The professor explained his intuition that, on average, tension measurement should increase going from left to right in the picture.

Mick needed to get permission from James, another mid-level manager, to have the lab techs collect the data. After the professor explained his proposal to James, the latter was blunt. "Tension measurements don't depend on how thread hangs from the spool. It can't make a difference. The experiment would be a waste of the lab tech time." Mick had the feeling that the outsider still needed to gain credibility. He was just a stat guy. What the heck did he know about manufacturing thread? However, Gregg's persuasion came into play and the experiment was conducted. The data, tweaked, rescaled, then modified, then further disguised, were, to some extent, similar to something similar to what is shown below[xii].

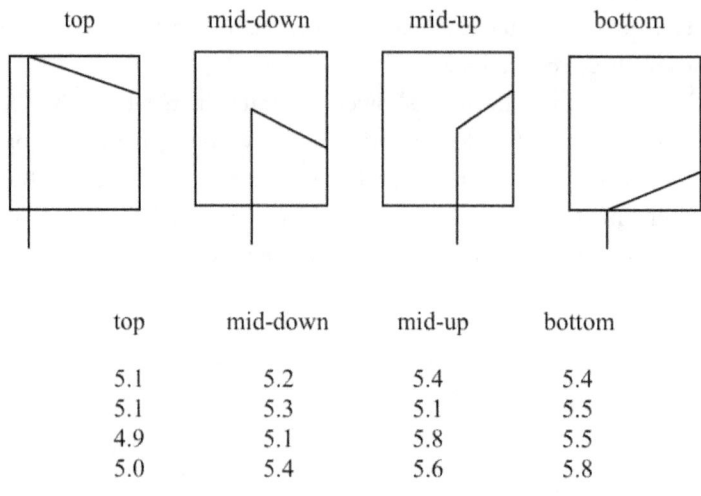

top	mid-down	mid-up	bottom
5.1	5.2	5.4	5.4
5.1	5.3	5.1	5.5
4.9	5.1	5.8	5.5
5.0	5.4	5.6	5.8

"See? I told you it wouldn't make a difference," James pronounced as he glanced at the data with Mick. The data collection had just been completed, and James had happened by just as the professor dropped by the lab to pick up a sheet that had the 16 measurements. "Look at that. The numbers are pretty much all the same. In fact, do you see that three of the mid-down numbers are greater than the 5.1 in the mid-up column? And there are two 5.1s in the top column and only one 5.1 in the mid-up column. I don't see that there is a difference in the four sets of numbers."

"You can't cherry pick particular numbers and compare that way. If one scenario requires only slightly more tension than another, you should expect to find examples like the ones you just spotted. If the trees in Forest A are generally taller than the trees in Forest B, that doesn't mean that every tree in Forest A has to be taller than every tree in Forest B. Actually, we even get a situation like that here in this data set. See the bottom and top columns? The numbers in the two columns don't differ by much, but every number in the rightmost column is bigger than every number in the leftmost column.

"I thought you weren't supposed to cherry pick."

"Sure. But you look at tip-offs to general patterns, not at individual numbers. I agree that the numbers in those two columns don't differ a lot, but the numbers in a single column are a lot more alike than the mix of numbers you get when you combine two columns. Statistics takes into account both variation within columns and variation across columns. The test that is used to compare these four scenarios is even called an analysis of variance. We'll now get to see if there is a sign that there are differences between the four forests. I was just heading to my office to do the test. I'll show you the results when I get them."

Explaining what goes on in an analysis of variance is far too intricate to describe here. So we herewith avoid discussing topics that make statisticians mouths water, such as normality and homoscedasticity. The spell checker faithfully pointed out to the author when he typed this paragraph that homoscedasticity was misspelled. He looked and found that, indeed, he had typed ec instead of ce. Mick dutifully thanked the computer for being sharp enough to catch the misspelling. And the reader can surmise what happened when the author transposed the letters into the correct order.

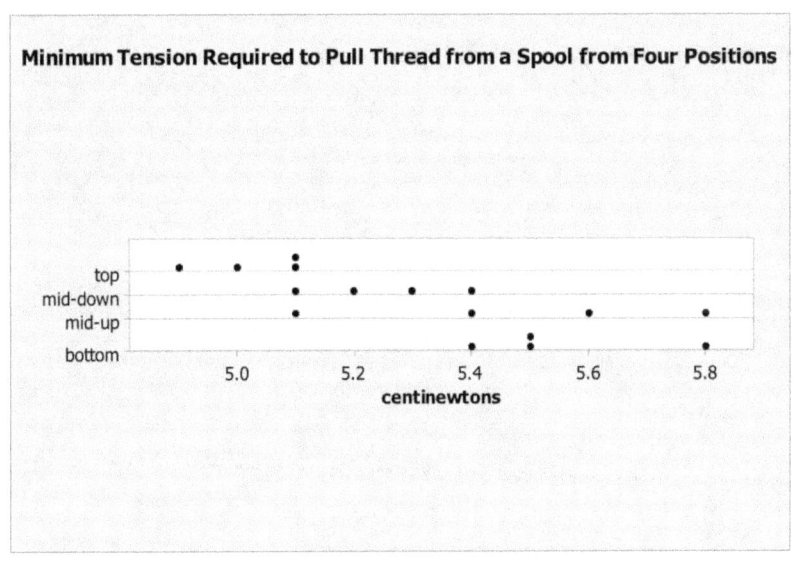

Although statistical background information is avoided, it may be helpful for the reader to refer to the plot of the data shown above, obtained using Minitab® Statistical Software[xiii]. For some strange reason, this type of plot is called a dotplot. Visually, the first things a statistician would notice are location and spread. On the preceding page, notice that the four hang-from-the-top dots are located near each other on the left side of the plot (where tension readings are low), the four hang-from-the-bottom dots are located near each other on the right side of the plot, the mid-down dots are near each other in the left-center of the picture, and the mid-up dots fall mostly in the right half of the picture and are a bit more spread out.

If thread position truly had no impact on tension, one would expect, ideally, the four sets of dots to be similar in location and spread – i.e., would all be similar in appearance. This is illustrated in the next dotplot. To decide whether thread position makes a difference, a statistician examines whether the point fallout of the four sets of points differs unusually from the kind of fallout that would be expected when thread position has no effect on tension. This is something that can be measured. This measure of unusualness is called the p-value of the test. The p-value for the analysis of variance on the data was .009, meaning that if position did not affect tension, one would expect only nine experiments in 1000 to have the four sets of points be as dissimilar as they were, or worse.

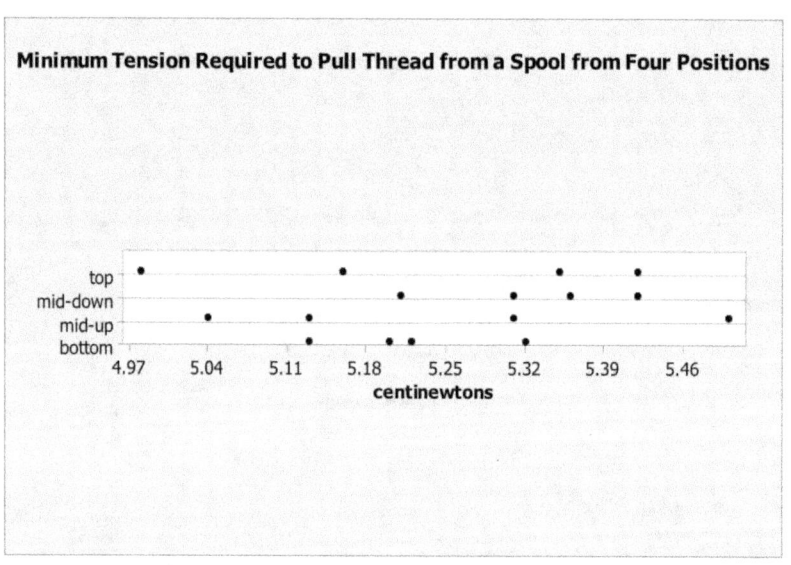

In statistics, it is common to conclude that the dependent variable (*tension*) varies significantly between treatments (*thread positions*) when the p-value of the test is .05 or less. Mick took the computer output and went to see James.

"You're saying that if thread position made no difference, the four sets of numbers could be as different as they were, or more so, by pure chance, but that chance is less than one time in 100 .... What was it? .009?"

"Yes. The data indicate that tension is affected by thread position."

"Well, thread position doesn't make a difference. We've never noticed that it did before."

What to do? thought Mick. James had been cold ever since the professor had first come to the plant. Mick figured that James must have viewed him as some kind of job threat. James had a role in interpreting much of the data generated by the testing lab. Part of his routine was examining data stored in spreadsheets, comparing current data to historical data in an effort to detect unwanted changes in quality. Feeling that his year would be easier if he could win James over, Mick had gotten in the habit of dropping by to tell him what he was working on and share his statistical observations. Actually, Mick had detected some gradual thawing in recent days. Oh yes, and on several occasions he mentioned that he was glad to be there to try to *help* and that he was hoping to do some good before his year was up. There was a big emphasis on *year*.

"This experiment was one of those one in one hundred – an anomaly. Repeat the experiment to see if the results can be confirmed."

Mick, perhaps overconfident from the *p*-value of .009, was not really worried about how a retest would turn out. It was just that the people involved with the tests were so close to the process, they didn't question its routines. Recognizing how drag might affect the lab's tension measurements was done by fresh eyes with no dog in the fight. Of course, eyes don't have dogs. Prizes for bad writing have been won for sentences like the preceding one. Working the envelope just for the heck of it is fun. Readers who have made it this far are allowed to roll their dogless eyes.

Actually, the professor was rather surprised by the small *p*-value of .009. Even with only four observations per thread position, the test was strongly indicating that thread position made a difference. *On the other hand*, he thought, *maybe I do need to worry. The evidence is strong that thread position makes a difference, which means the next p-value should also be small. But the p-value depends on the data, which vary somewhat from experiment to experiment. What if the next p-value turns out to be, say .07, which is suggestive but falls just short of being significant? James might decide that the results of the first experiment are not confirmed.*

In James' defense, tension measurements were not a life or death issue. The spools they were producing at the time met product specifications.

However, the professor also was devoted to the old quality mantra, the one that said that merely meeting specs was never good enough. If you improved the process, you would make a more consistent product, one with more chance of being robust enough to be affected minimally if something went wrong with the process temporarily. Such robustness would increase the chances that the product produced in the interim would still meet specs and/or be sellable. Also in James' defense, no company that makes a successful product would, or should, replace a standard operating procedure that works, unless there is rock-solid evidence, a term for which there is no generally recognized definition.

The p-value of the next experiment was .03. Mick was happy when he went to see James.

"I think you've got another anomaly," James pronounced.

"Well James, suppose you think different thread positions might affect tension, so you do an experiment in which your threshold for declaring significant differences is unusualness to the usual tune of one chance in twenty. Then suppose you get significant differences on two experiments in a row. If thread position really doesn't make a difference, the chances of that happening are only one in 400. The evidence is really strong here that thread position matters."

"Well, the kind of tension measurements you get can vary a lot from spool to spool. Differences between spools could explain the differences you're seeing."

"Actually, I am aware of that. In each of the two experiments, all 16 measurements came from one spool. That keeps any differences that may exist between spools from leading to an incorrect conclusion. Any significant differences found can then to be attributed to thread position, not to differences between spools."

"OK. But even with one spool, there could be issues. Just as denier varies a little from point to point on a line of thread, you can have quirky sections of thread on a spool. Tension measurements are sometimes consistently higher on one section of thread than on another somewhere else on the spool."

"I think we're good there, too. Rather than getting, say, four "mid-up" measurements in a row from the same general location on the thread line, I had the lab tech obtain measurements in an order I prescribed. So for example, from a given section of thread, you get a top, a mid-down, a mid-up, and a bottom measurement, not four "mid-up" measurements. That guards against one thread position being made to look different than the others as a result of its four values all having come from the same quirky section of thread. The four thread positions would all be impacted in the same way by a quirky section of thread."

Mick liked the way James was smiling at him.

"Hey, it's what I do," the professor said as he smiled back.

"OK. Do the test one more time just to confirm the results."

Mick knew the war would be won by winning the next battle, but he didn't want to take any risks. Since there would be more chance of detecting significant differences if the experiment used larger sample sizes, the next time a sample of spools was taken to the lab, he had the lab tech obtain six observations from each of the four thread positions. A sanitized version of the data is portrayed in the dotplot shown below.[xiv] MINITAB, the software package that did the analysis, announced the p-value of the experiment in its usual fashion, rounding to three decimal places: .000.

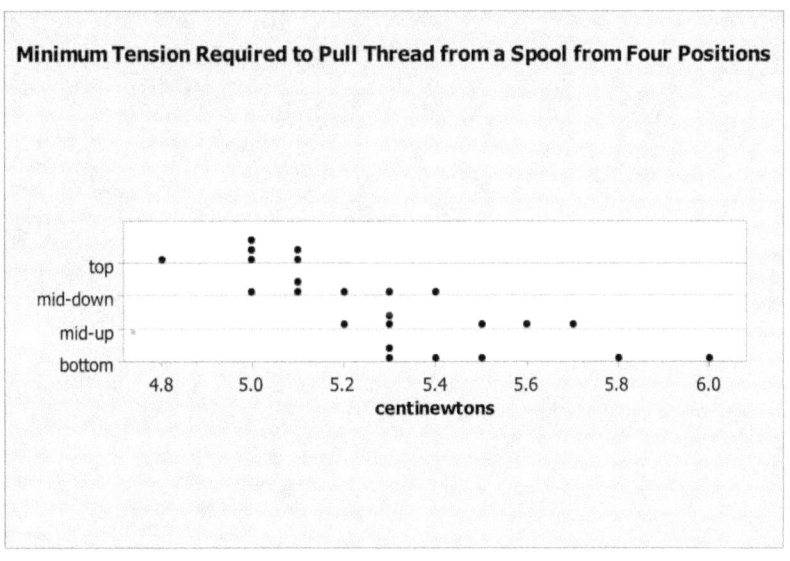

At that point, upper management was convinced. Lab techs were all asked if they had a preference as to which of the four methods should be adopted as the official method. Based on the mechanics of obtaining measurements, *hang from the top* was their unanimous choice. From that point on, *hang from the top* became the standard operating procedure for the tension test.

There was a lot about which to be satisfied as a result of the experiment. Lab techs were now producing more meaningful numbers because they were performing the tension test the same way every time. Also, instead of test results varying widely because thread could hang from an infinity of positions between the top and bottom of a spool, there was less spread in the values the lab techs were recording – i.e., precision was increased. And finally, the reduction in measurement spread meant it was much easier to get the same number at least twice in three replications – i.e., the average amount of time to do the tension test on a spool was reduced. Mick had helped the company do something a little bit better, and it gave him a high. It was one of a number of highs he would experience that year.

31 CHEERS, TREY

It was three weeks into the spring semester. One MATH 104 class that the professor was teaching reminded him about what he considered to be a serious problem facing the College – the percentage of freshmen who didn't know what their responsibilities were in the learning process. It had been gradually increasing for years. If we could get into their heads, he mused, before they set foot on campus, let them know how much effort they would need to put forth, that they would need to stay on top of homework every day, to prepare for exams earlier than the night before the exam, that a final exam would warrant serious preparation, and so on, we could save many students from themselves. The result would be fewer D and F grades, fewer students put on academic probation, fewer students who didn't finish in four years, and fewer students who lost scholarships.

He was teaching two sections of MATH 104. One section was typical – students with a mix of math ability levels and who were attentive and who asked good questions. They also were pleasant to interact with. The other section had three annoying students. They would tune in and out repeatedly during class, and upon tuning in, each was likely to ask a question – typically one that would have been good 15 minutes earlier. If the question had a one-sentence answer, Mick would answer the question just to keep things moving. Otherwise, he'd invite the student to come by after class or to come to his office. None of the three had ever followed through on that offer.

One Monday, the professor made an announcement before going over homework problems.

"I have noticed that this class covers a few minutes less material than the other class – nearly every day. The primary cause is a few people who tune in and out repeatedly during class and ask inappropriate questions – about issues that were addressed earlier. I am happy to answer questions, and I actually encourage it. But we cannot take everyone's time to personalize the course to people who don't pay attention. So here's the deal. If you have not paid attention during part of the class, you CAN'T ask a question in class. Write it down and I will answer it after class or during office hours. I will be happy to answer your question, but we can't take up everyone else's time to personalize the course to you." Mick figured that the appropriate students would recognize themselves as the targets of the message.

Trey Wheeler was the most chronic offender. From the front of the room, Mick could see student faces, and it was clear that Trey hardly ever tuned in during class. On the Friday after the professor had made the announcement about asking questions, Trey asked two questions, and each time, the professor responded that the two of them should meet after class. Trey then asked a third such question, interrupting while Mick was explaining a concept.

"Trey, you need to listen and stay engaged," instructed the math professor forcefully, making eye contact before returning to the explanation. He hoped that Trey had gotten the message.

Mick wondered if the guy was ADD and not taking his Ritalin. When Mick finished explaining the concept, he looked for an indication that the students had got it, which he did by scanning the room to see if there were any puzzled faces. Trey's face was the only standout. Within seconds he blurted out a really bonehead question. It wasn't so much a question about the math as it was a question asked to take up time. *Was this guy playing games with me?* the professor asked himself.

"En GAGE," Mick instructed. He paused, fixing a stare at the student. It was an unmistakable signal to cut the, well, to get with the program. Embarrassing a student in front of the whole class was something he took no pleasure in, but the student had to lose that kind of contest.

The professor returned to the lesson and, within a few seconds, Trey picked up his skateboard and backpack and left the room in a snit. He had to walk across the front of the room, right in front of the professor, to get to the door. His face and body language were easy reads – hostility, anger, tantrum. As he passed the professor nose-to-nose, which was necessary due to the location of the front desk, he glared at Mick with those emotions. The professor's reaction when the student passed by him was to smile back pleasantly and say "Cheers" to the departing student.

The professor figured that Trey would drop the course. And he did, but not before coming to the next three classes, during which he was more attentive and asked occasional, better-thought-out questions.

Trey had a defender. Alec had an adolescent, Joe-Cool demeanor, perhaps because the material came easily to him. He was one of the other two who would tune in and out during class and still ask questions. Sometimes Alec's questions were insightful. Sometimes they were designed to show the other students that he was sharp. Sometimes they showed that he hadn't been paying attention. Sometimes, particularly near the end of the class period, they were stall tactics designed to keep Mick from reaching the point to assign homework (which never worked). Alec averaged one or two questions per class. On the day that Trey had stomped out, Alec came up after class.

"Trey was just asking questions. Sometimes students don't understand everything. We have to be able to ask questions," Alec implored.

"I agree," Mick responded. "But I can see something you can't – faces. Trey is one of several students who sometimes don't pay attention. We can't take everyone else's time to answer questions that wouldn't have been asked if the student had been paying attention. This isn't high school, you know."

"That's true," he agreed.

Although not as extreme as Trey, Alec and the female student who sat in front of him did qualify as EEEs (effete, empowered, exasperating). Besides occasionally coming out with inappropriate questions, they would sometimes whisper to each other and become a distraction. In almost every class meeting, the professor would need to shush them. On top of all of that, Alec was a whiner. All too often when he'd get a quiz or exam back, he would come up after class, show Mick one of the questions, and try to persuade the professor that he had been too harsh in deducting points.

One Monday late in the term, Mick returned the third exam. As usual, Alec came up after class to argue over two points that had been deducted on a 10-point problem. The question stated that an experiment had been done to determine if a new drug was better than a standard drug that was used to relieve chronic back pain. The p-value of the experiment was .004. [The reader should think of .004 as being: the probability that the new drug would, by pure chance, outperform a drug it's no better than by at least as much as it did.] The question was, Would the null hypothesis [that the new drug is no better than the standard drug] have been accepted or rejected? Alec's answer was that at either of the significance levels .05 or .01, the null hypothesis would have been rejected.

"I don't understand why I lost two points," the student said.

"If you had said that the null hypothesis would have been rejected at any reasonable significance level, you would have received full credit. As it is, you only lost a couple of points. What you said about those two significance levels is true. But remember how I told you guys to be on the lookout for extremely small p-values like that? .004 is less than any reasonable significance level people use. We've done examples like that before. I want you to think at that level. Next time you see a question like that, I bet you'll get it."

"But .004 is less than .05, and you told us that when a significance level is not specified, we should use .05. It's in my notes."

"I wouldn't have said that."

"It's in my notes. I'm trying to get the best grade I can in here and I don't know what I need to do to get higher exam grades."

"Write things that don't make me take off points."

Alec laughed derisively. Actually, Mick hadn't intended to be sarcastic, he was trying to be honest. But the honesty was colored by a semester's worth of whining that was wearing thin.

"I don't think taking off two points is fair," Alec began again, pausing to inhale as if he were about to describe a lengthy line of reasoning. Actually, he did start to say more, and the professor was about to cut him off.

"This is NOT high school," Melissa interrupted in the scolding tone of an old-time school marm. Melissa was a student who, like Alec, wanted to talk with the professor about her exam. She had been standing there waiting. Between Alec's whining and his classroom behavior during the semester, she had finally had her fill.

Her words of admonishment were the same words Mick had said to Alec when Alec had come to the defense of Trey – words to which Alec had agreed. It took a fellow student saying the words to get his attention. He appeared shocked.

"But we have to be able to ask questions," he continued. "And I always try to be polite."

"No you don't," the professor observed. "What about the day before the exam when I was starting class and you were talking to the girl in front of you? It was obvious that class was starting. Your conversation was very loud and went on long enough that it had to be obvious to you that you were competing. I had to yell 'HEY!' to get you to stop. What the hell goes on inside your head, anyway?"

"I am NOT an AARDVARK," Alec asserted vehemently, in response to what he must have perceived as an impromptu double-team. Actually, Alec did not say *aardvark*. He used a synonym. It wasn't just his words that were surprising, but the tone in which he delivered them. It was as if there had been an invisible presiding judge present who said, "Alec Smart, you are charged with being an aardvark. How do you plead?"

"Look," Mick said. "We only have four class meetings left. Study hard. Quit talking to the girl who sits in front of you. And pay attention. If you get an A, that's fine with me. Any grade you get is fine with me. Now go do."

On the walk back to his office after having helped Melissa, the professor tried to imagine what kind of citizen Alec would be during the last few days of class. He turned different scenarios over in his mind. Had he gotten the message? When Mick returned to his office, he was surprised to see Alec waiting at the door.

"I really am a nice guy. It bothers me that any students may view me as a jerk. I'm not. Does anybody think that?"

"You can probably win over some minds if, in class, you do the things I just talked about. The two of you need to quit whispering and pay attention. Then you won't wind up asking questions that waste other people's time."

And indeed, Alec truly was a better citizen in the classroom those last few days.

32 ON PLAYING JURY POOL

It had been more than fifteen years since the professor had received a summons to report for jury duty. And now he had received another. He thought about that first experience, when he actually had been drawn from the jury pool to hear a case. The first thing he would always recall when thinking about that trial was the gutsy juror who'd had enough of a dislikable attorney's antics to say something from the jury box.

A key piece of evidence in that case had been the blood-alcohol level of the driver at the time of the accident. Because he had been taken to the hospital by EMS, his blood-alcohol level had not been measured until fifty minutes after the accident. A physician who was an expert witness for the state had testified that a standard extrapolation of the Breathalyzer reading taken at the hospital indicated that the motorcyclist's blood-alcohol level would have been .127 at the time of the accident. The young man's attorney had done everything in his power to deprecate the credentials of the expert witness and the trustworthiness of the extrapolation. The value .127 had been tossed around so much on the first day of the trial that it was embedded in every juror's brain.

Likewise embedded was the expert's hourly fee. "Are you paid for your testimony?" the attorney had asked. The professor-juror had immediately recognized by the phrasing the word-game that was being played. *I am paid for my time, not for what I say*, is what Mick was hoping the response would be. But the physician simply stated his hourly fee. Later in the jury room, several jurors had expressed open disgust that the fee was so high. The professor chose to not offer up that he had ever served as an expert himself.

Consequently, he also hadn't pointed out that the fee didn't seem out of line, in so far as expert or attorney fees went. He was prepared to say something had he needed to, but the fee seemed to have no impact on the jury's deliberation process. In fact, the juror most outraged by the fee offered up a yin-to-the-yang observation: "I know that bar he was at before he drove away. He was as drunk as a skunk."

The driver's attorney did nothing to improve the image that some have about lawyers being unprincipled lowlifes. This guy could have been sent by central casting (or by the World Wrestling Federation) to be a villain. On day two of the trial, while the attorney was making his closing remarks, he again started badmouthing the expert's testimony. He had no difficulty remembering the expert's hourly fee and describing it in class warfare language, but feigned the inability to remember the value of the physician's extrapolation.

"He came up with some number that is supposedly not subject to question. A number. What was that number?" he asked himself searchingly. He walked over to his table and rummaged through some papers. He then walked around, seemingly in deep thought, trying to recall the number. The multitasking – walking and thinking – was impressive, but not in the way intended. "Well," he dismissed, "the *expert* came up with some number that is not supposed to be subject to question." The attorney had clearly finished talking about *the number*.

A loud "point one-two-seven" broke the silence. One of the men on the jury had just done what the entire jury wanted to do in reaction to having had to put up with this guy for more than a day. The juror's delivery had a tone of offering assistance, but also irritation. *Your acting is atrocious. And oh yeah, do you think we're idiots?*

"DON'T ... DO ... THAT ... AGAIN," scolded the judge, who stared daggers at the offending juror. The pauses between words conveyed the desired fear-of-God impact. *But inside, we all were smiling,* Mick recalled.

Will I serve on a jury this time? he mused, thinking about his peek, years earlier, into one court system's process for selecting jurors from a jury pool. At an attorney's request, he had gone to the attorney's office to discuss the possibility of serving as an expert witness in statistics. This was one of those *first visits* with a potential client. There was no charge. The professor used such meetings to size up a case and determine whether he could help.

The case involved a lawsuit in which it was alleged that a process of selecting jurors from jury pools was biased against minorities. Salient background information was that blacks made up about 33% of the population, and also about 33% of all of the people called to serve in jury pools for the last two years.

"The problem," the attorney maintained, "is that on many juries of twelve people, there are less than four blacks on the jury."

"What is the process for selecting jurors from the pool?"

As she explained the process mechanics, it became clear that every member of the jury pool had the same chance of being chosen to be on a jury. Moreover, an obvious consequence of the process mechanics was that every possible mix of twelve people had the same chance of being the jury that would be chosen. In mathematical terms, the latter implied the former, but not the other way around, a difference that someone in statistics would notice. Basically, there was no mathematical opening with which to criticize the procedure.

The mathematically oriented reader may wonder about the difference between the two scenarios. To illustrate with a small-scale example, suppose a jury of two people is to be chosen from a pool of four individuals who are named A, B, C and D. Suppose only two possible juries are considered: Jury #1, which consists of A and B, and Jury #2, which consists of C and D. If a coin toss decides which of the two juries hears the case at hand, each of the four individuals in the pool has the same chance (fifty-fifty, or probability ½) of being on the jury that hears the case. Of course, A and C cannot simultaneously be on the jury that hears the case, but they individually have the same chance of being on the jury that does. On the other hand, suppose a four-card deck – with cards labeled A, B, C and D respectively – is well shuffled. If two cards are dealt and used to identify the makeup of the jury, then each possible pair of people – AB, AC, AD, BC, BD, CD – has the same chance of being the jury that is chosen. Each of the four people has three chances in six, i.e., a fifty-fifty chance) of being on the jury that is chosen.

"The process you just described is a fair one. Blacks are not treated differently," said the professor.

"Yes, but many twelve-person juries have less than four blacks." The attorney was clearly dismayed with that fact.

"OK. But there also will be juries with more than four blacks."

"True. But what about the juries with three, two, one, or even no minorities? That's just not right."

"The process you described selects jurors at random. The average outcome of the process you described is four blacks per jury. Many juries will have exactly four, some will have more, some will have less. That's a mathematical consequence of being a random process."

"To my way of thinking, the ideal process would have the average be four, but still make sure that every jury of twelve has at least four minorities."

"No random process that has an average of four would put at least four on every jury. The only way to have the average be four and guarantee that at least four would be on every jury would be to impose a rule that every jury has to have exactly four blacks on it. But then the process wouldn't be random any more."

The attorney's face showed that this was not what she wanted to hear. It was time to bow out. The professor told her that he wasn't sure he could help her, also that he was really busy lately and didn't think he should be taking on a case. She had to have read between the lines, but effervesced graciousness and understanding in her farewell.

Thus had been the totality of Mick's acquaintance with juries at the time he reported to the courthouse to comply with the recent summons. After passing through a security checkpoint at the entrance and taking an elevator ride, he was greeted by a gauntlet of bailiffs, each immaculately dressed in a coat and tie, each with a professional demeanor that hinted of a past stint in the military or as a police officer. The first bailiff asked for the professor's name and, looking to put a check mark by a name, proceeded to thumb through a multi-page computer printout. He came up empty.

"Could it be that you were called to serve at an earlier date and had your service postponed until today?"

"Yes, that's true. I was supposed to serve last week, which posed a difficulty. I called the clerk of court, explained my situation, and she was kind enough to reschedule me."

The bailiff defaulted to a sheet of copier-paper that had about twenty names on it, found the name, wrote a check mark by it and, still glancing at the page, informed Mick that he was juror number 265. The professor continued through the gauntlet, receiving a *free-parking* pass to use at the parking garage and a large blue circular sticker that said "JUROR," which he was to wear on the front of his shirt. He was ten minutes early when he entered the room and sat down to occupy one in a vast array of chairs.

When nine am arrived, he guesstimated that there had to have been close to ninety people seated in the jury pool chairs, a yield, he would later find out, that resulted from close to three hundred summonses having been sent out. Each potential juror present was to stand when called, give name, juror number, age, occupation and where he or she worked, and spouse's occupation. Retired people were to identify what their last job was and for whom they worked. Not many in the pool said they were single. One of these was a woman who announced, assertively, that she was divorced. Another announced, with great reluctance, that she was a widow. There was a vast cross-section of humanity age-wise and race-wise. Also occupation-wise. There were clerks, cleaning ladies, attorneys, nannies, unemployed people, an assistant to the mayor of a small town, a fireman, a correctional institution guard, construction workers and foremen, housewives, college professors, self-employed people with a variety of interesting lines of work, and many more.

Seated against the only wall with windows were attorneys who would be trying cases that week. They were not introduced by name. Each of them had been provided with a list containing juror names and numbers. As each jury pool member rose and provided the requisite information, the attorneys made jottings on their lists. The professor had the brief sensation of having entered a beauty contest, but quickly changed his mind. Instead he felt that he was more like a widget on a manufacturer's conveyor belt, being examined by a quality inspector who was looking for defective widgets to remove from the product stream.

After a bathroom break, the gathering was presided over by a judge who stood at a podium at the front of the room. She asked a series of questions designed to identify people who needed to be excluded from the jury pool. Had anyone ever served time as a convicted felon? After a moment of silence, the judge proclaimed, "There are none." Was there anyone who was 65 or older and wished to be excluded? Again after a pause the judge said, "There are none." Was there anyone who could not read or speak English? Again there were none, although there might have been some who would have stood up to speak if only they understood English a little better. Altogether there were, to be precise, oodles of such questions. Occasionally, someone would stand and respond to a question with the particulars of his or her situation, and the judge would dismiss the person from the jury pool.

The judge had an assistant who was seated at a table near the judge's podium. On the table, arranged into neat rows and columns, were slips of paper a little larger than business cards. The goings-on suggested that each slip had a juror's number and name on it, one slip for each member of the jury pool. Whenever the judge announced that a juror was dismissed, the juror would be asked to identify his or her juror number, after which the assistant would take a quick glance at the array and, without hesitation, remove one slip from the array. Ultimately, a handful of people were dismissed, including one of the two who responded to the invitation to speak with the judge in private to discuss why they thought they should be dismissed.

Also on the table was a device that one might call a lottery bin, a container made of wood. It had two opposing vertical sides that were octagonal in shape. The bin was connected to a wooden base by a metal frame, and a handle at the center of one of the octagonal sides allowed the assistant to make the bin revolve on its axis like a Ferris wheel. The assistant turned the handle to bring one of the eight *outside* sides, the one with the door, to directly in front of her. After opening the door, she used her hands as if she were a poker player raking in chips from a pot. First she swiped together the slips of paper that were closest to her, creating one fistful, which she placed in the bin. She swiped her hands along the tabletop once again to obtain a second fistful of slips and placed them in the bin. And so on. Once all of the slips were in the bin, she closed the door and turned the handle to make the bin revolve.

The professor was struck by two facts: the slips had been put into the bin in bunches; and there did not appear to be a lot of air space left in the bin. Not much mixing was going on as she turned the handle at a gentle pace.

Instead of slips of paper, if they're not going to use software, they'd be better off using cards and shuffling them. Mick's thoughts turned to Perci Diaconis, the well-known mathematician/statistician who had become even more famous – in mathematical circles, anyway – for his having presented a convincing mathematical argument that giving an ordinary 52-card deck seven good shuffles would satisfy just about anyone's practical criteria for what it means to put the cards in random order.

The assistant again brought the bin door to directly in front of her and opened it. Because two trials were to be held that day and there were 80 pieces of paper in the bin, she would draw out forty to identify the individuals who would be sent to a particular courtroom. It is left to the mathematical reasoning of the reader to deduce which jurors would be escorted to the other courtroom. Instead of pulling out one slip at a time, she reached into the bin with one hand, pulled out a fistful of slips and dumped them in a pile on the table. Then she drew out another fistful. And so on. It was obvious that the assistant had a routine. She stopped twice to count how many slips were already drawn, then drew several more slips individually to bring the activity to a close.

"Would the jurors with the following numbers please stand in line against that wall by the bailiff." What Juror number 265 heard next went something like this: Juror number 271, Juror number 258, Juror number 267, Juror number 9, Juror number 8, Juror number 264, Juror number 5, Juror number 12, Juror number 269, Juror number 263, Juror number 17, and so on. *At this rate, how can they miss me?* Mick thought. And in due course, he joined the line.

The mental gears were turning. *Wow! Input in bunches, output in bunches. OK, sure, there are a few numbers like 37 or 118. But people numbered 20 and below were selected in droves. Likewise, people in the mid-200s – ones whose jury duty had been postponed – are clearly overrepresented ... for this trial anyway. Overrepresentation of this latter group might have negative consequences. If a jury were flooded with such people, how might it affect the decision-making process? Do these folks have consistent commonalities? If so, what characteristics might they be expected to have?*

Mick didn't remember there being many – or any – teachers among the eighty. That made sense, because most K-12 teachers would be busy for a few more weeks. Teachers would have requested a deferral until the summer. His next thought was a surmise about the people numbered in the 200s: *busy people ... active people ... people who do things. What kinds of occupations or personality types might that be?*

Mick's group of forty was escorted to a courtroom and seated in the observer section. Within a few moments, a bailiff announced, "All rise. Court is now in session." A judge entered the room, told everyone to be seated, and announced that the case to be tried – officially referred to as *So-and-so versus So-and-so* [they were not related to each other] – was a lawsuit involving personal injury and an automobile accident. The judge then asked the two sets of attorneys to introduce themselves, identify the firms they worked for, and introduce all of the individuals seated with them at their respective tables.

When this activity was done, the judge asked a set of questions designed to cull people from the jury pool for this case. Was anyone related to or a good friend of either the plaintiff or defendant? Was anyone related to or a good friend of any of the attorneys? And so on.

To the question "Has anyone had business dealings with or been represented by either law firm?" eight (or so) people stood up. Two had been law firm clients. The others were attorneys whose firms had had professional interactions. The judge asked each of the eight-or-so if the past or current interaction would keep the him or her from being fair or from applying the law as it related to case at hand. Seven, without hesitation, said that it wouldn't. The long silence and body language of the eighth, an attorney, reminded Mick of the joke Jack Benny used to tell about the robber with a gun who had threatened him by saying "Your money or your life." This attorney could similarly have said, "I'm THINKING. I'm THINKING." In the end, though, he responded as the others had, but grudgingly and with apparent pain. *Was this a signal to a quality inspector?*

At some point the judge asked if anyone in the pool had been a party to, or had a relative who had been a party to, a court case that involved bodily injury. Lots of people stood up, including the professor. Each person got to explain the nature and cause of the injury, whether the association was with the plaintiff or defendant side of the case, and whether the case was still pending. All of these jurors responded in the negative when asked if the experience would keep them from being fair or from applying the law to the case.

The jury pool was excused from the courtroom for what the professor assumed was a bathroom break. After returning from the break, the forty learned that ten of them had been chosen to compose a new pool. Six members of this new pool would be the jury that would hear the case. *Why didn't they pick the ten in an open, transparent fashion,* he thought. *No lottery bin is in sight. What was their process?*

The forty jurors then listened as each of the opposing attorneys struck two people from the pool of ten, which they accomplished by taking turns at striking one juror at a time. The ten jurors had been renumbered: juror number one, juror number two, ... , juror number 10. This made it impossible for the professor to know if he was struck from the group of ten, or even if he was in the group of ten. *Why was this process not open?*

Ultimately, six jurors were called out by name and asked to report to the jury box. Those not called, including the professor, were dismissed for the day.

It was a similar story the next day, except that twenty people were chosen in order to obtain a jury of twelve, and each attorney was given four strikes. On the second day, when the twelve jurors entered the jury box, again sans Mick, and the judge formally asked each attorney to look at the jury that had been selected and state if there was any reason that this particular jury would not be acceptable, everyone in the room except the always-serious bailiffs chuckled when one of the attorneys wisecracked that he wished he'd had more strikes.

So whom would attorneys want to strike from a jury pool? One group might be fellow attorneys they might expect to work with or against in future cases. Mick recalled the time when his wife had wanted to take a graduate course in nonparametric statistics, and he had happened to be the one teaching the class at the time. So Libby merely *sat in on* the course instead of enrolling. No good could ever come from having to assign one's spouse a grade. What spouse would ever want to set up a potential catastrophe like that? *Would trial attorneys often be in the group of those whose jury duty was postponed?*

The professor also suspected that in personal injury cases, an attorney wouldn't want on the jury people who knew much about such cases, particularly anyone with experience at watching how someone got shafted. Absolutely, plaintiff lawyers would want to strike people who may have acquired firsthand experience at recognizing ambulance-chasing lawyers or scamming chiropractors.

So the professor was torn. On the one hand, he wanted to be a good Samaritan, call the clerk of court, and explain why a frequent consequence of how they used the lottery bin would be an overrepresentation (or underrepresentation) of people whose jury duty had been postponed. On the other hand, he had a philosophical aversion to having attorneys exercise a lot of power in shaping juries. He was pleased when that attorney said he'd wished he'd had more strikes. *What to do?*

In the end, the Samaritan tried. An official familiar with the mechanics of how juries were determined had responded to his email with a phone call. The man answered the questions about *process* that Mick had posed in the email. However, when the professor pointed out, as he had done in the email, what he did for a living, and asked, "Are you open to input?" the man knee-jerked, "We follow state law." It was a no-brainer to recognize a contextual synonym for "No."

Nonetheless, reasoning that he had not specifically heard the word *no*, Mick crusaded on, announcing that he felt an obligation to share some observations. He explained how their fistfuls-in/fistfuls-out procedure had resulted in an overrepresentation in the jury pool of people whose jury duty had been postponed.

"We always welcome juror feedback about how we can improve our procedures." *Aha! Another contextual synonym.* Helpfully, however, the man was quick to point out that the court would go to a computerized selection process in a couple of weeks. *I wonder how he'd respond if I offered to check it. Nah, better not.*

33 YOU NEVER KNOW HOW THEY'LL QUOTE YOU

His phone was ringing. It was Pat, calling from the math department office.

"I've got someone on the line from the Charleston City Paper. Could you help this person?" she queried.

"Aha. The media with a math question. Sure, put them on."

In the two seconds before being connected with the caller, the professor had two flashbacks. He knew he had to be careful answering human-interest questions from the media, who were often more interested in brevity than getting the math right. The first incident Mick remembered was the call from that guy at a local TV station. It was on a Friday the thirteenth, and the station's two news anchors were going to do a brief back-and-forth about superstition on the evening news. What they wanted from Mick was the answer to a question. "What is the probability that today is Friday, the thirteenth?" the guy asked. The professor remembered thinking, but of course not saying, *One. Bet the house on it. It's a sure thing.* Like inside jokes everywhere, wisecracking about how a probability question was phrased would be appropriate only among math people. After some discussion, it was clear that what the guy really was trying to ask was, *How many times a year do they come?* Mick had worked it out and called back with the answer and an explanation for why it depended on the year (the explanation involved weekly cycles, monthly cycles, and special cycles, such as leap year, all operating simultaneously). And the next day, the professor got a call from a *gotcha guy* informing him that he was wrong. The gotcha guy taught at a two-year college. "I agree they were wrong. But I wasn't," Mick had responded. "They left out two key words that I told them to be sure to say: *on average.*"

And the *gotcha guy* got it right away. He even became sympathetic, possibly because such things had happened to him, too.

The other incident Mick remembered was having called a reporter from a local newspaper to ask why someone had changed the wording to which that same reporter had promised to be faithful. The answer was that they had shortened it a little to make it simpler. Not only was it simpler, it was also flat-out wrong. Once the media had two strikes on him, he had focused even harder on guarding the plate.

"This is Mick Norton. How can I help you?" The caller's question was about one of the Spoleto (annual art festival) events. You never knew what you would get from such calls. Happily, the professor had no serious issues with the outcome of this call:

Charleston City Paper, May 24, 2006[xv]

OPERA & THEATRE 52 Pick Up

According to College of Charleston mathematics professor Mick Norton, the chances of someone seeing the same exact version of *52 Pick Up* that they saw the last time it was performed in Charleston is less than one in 80,000,000,000,000,000,000,000,000,000,000,000,000,000,000,000 ,000,000,000,000,000,000,000,000.

That number — eight followed by 67 zeros — is so fantastically large it doesn't even have a name, according to Dr. Norton. (If you really want to waste your time, multiply 52 times 51 times 50, and so on, all the way through to 1, to get the same answer.)

Here's why: the play dissects the beginning, rise, fall, re-rise, and final failure of a love affair into 52 short scenes. Each scene relates to a card in a deck. The actors Llysa Holland and Andrew Litzky present the deck of cards to the audience at the beginning of the play, and then throw the cards up into the air and randomly pick up each scene from the floor — thus beginning the world's most elaborate "game" of 52 pick up. ...

34 PINK AIN'T MY COLOR

Some professors in the math lunch group were eating at a table in the faculty lounge. The topic of the moment was a movie about singer Johnny Cash, *I Walk the Line*. The discussion touched on topics in some of his songs and his life as portrayed in the movie. Mick's contribution had been about Cash's famous performance for inmates at Folsom Prison. A colleague from the English department entered the lounge, sat down at the table and joined the discussion. The English professor began to laud a movie he had seen not long before, one about singer Ray Charles. For a time, the group discussed the two singers. Part of the fun, in addition to being a necessity, was explaining some of what was discussed to Dinesh, the math colleague who hailed from India.

Shortly after lunch, an attorney from Greenville, South Carolina called the math department and asked for Mick. After talking with the caller, the professor wondered if this would be the only day in his life in which the topic of prisons would crop up twice. A Department of Corrections (hereafter referred to as DOC) was in need of someone who could analyze data.

It seems that DOC officials had noticed that in the last few years, the number of sexual assault and sexual misconduct infractions committed by inmates against staff or against other inmates had been on the rise. The caller, an attorney representing the state, explained the difference between sexual assault and sexual misconduct, and gave illustrations. Some of the illustrations were pretty gross and shall go unrepeated. Suffice it to say that assault generally involved someone being the recipient of inappropriate contact, whereas misconduct generally did not involve contact.

DOC had a comprehensive list of transgressions that inmates were to avoid. Whenever an inmate was charged with committing one of them, there was a judicial system internal to DOC that would find an accused inmate either guilty or not guilty. When an inmate was found guilty of committing sexual assault or sexual misconduct, he or she would do time in lock-up (think *solitary*) and lose some visiting privileges. Of late, however, with the number of instances increasing at a seemingly alarming rate, officials felt that they had to do something more. Their attempt at a solution was to add to the usual punishments a requirement that when the inmate got out of lock-up and was returned to the general prison population, he or she would have to wear a pink jumpsuit for a certain period of time. A justification given was that individuals wearing jumpsuits of a different color would stand out, making it easier for guards and other staff to know which inmates deserved keeping a little extra eye-on. The pink jumpsuit policy had been put in place on March 1, 2005. Mick's first thought was that pink, out of whatever possible colors were possible in the jumpsuit business, was an interesting choice. *What inmate would want to wear pink?*

One offender, upon being released from lock-up, refused to wear a pink jumpsuit. He sued DOC, alleging that wearing the suit would put him at increased risk of being attacked.

And so, on a bright November day in 2006, the professor took a trip to a state prison to meet with several DOC officials. Three statistics questions were discussed at the meeting: A) Would the data readily available from routine reports make a compelling enough case that the occurrence of sexual assaults and misconduct offenses had risen dramatically in the last few years? B) Would data show that the pink jumpsuit policy was having a deterrent effect? and C) Were people more at risk of attack after donning a pink jumpsuit than they were before?

The professor found that getting at the data needed in order to analyze question C) was, logistically speaking, a royal pain. One database had a record of which inmates had been found guilty of offenses committed within the prison system, but it didn't clarify the nature of the offense. By searching a second database with the inmate number of someone found guilty, the incident report that had led to charges could be found. Each such incident report would then need to be read by a human being to determine whether what the inmate had done would require the inmate to wear a pink jumpsuit. And finally, for each inmate who had donned a pink jumpsuit, incident reports would need to be searched in order to find all occasions when the inmate had been victimized, so that before and after (wearing the jumpsuit) data could be compared. Yes indeed, collecting the data was a royal pain. Hoping to be pleasantly surprised, Mick took a tentative stab at it and decided that the task was too daunting. Say hello to applying statistics in the real world. When he told DOC attorneys about the problems, they told him that he didn't need to continue to work on that particular question, that if the other side wanted to go to the trouble of making a statistical case that inmates were more at risk after wearing pink, that would be their case to make.

As to question B), the data that were collected gave indications that the pink jumpsuit policy was beginning to have a deterrent effect, at least in the sense that graphs were suggestive. However, while there was plenty of *before*-data, the new policy hadn't been in place long enough for there to be much *after*-data, certainly not enough to have a chance of finding a significant difference as the result of a statistical test. Being able to say that data were suggestive or promising was a far cry from being able to say that the pink jumpsuit policy was making a statistically significant difference. Say hello again to applying statistics in the real world.

Answering question A) in the affirmative – that is, making a case that sexual assaults and misconduct offenses had risen dramatically in the last few years – was easy. As the old saying goes, a picture is worth 1000 words. In a legal setting, a good graph could be more convincing than the well-explained results of a statistical test, even if the test findings were significant. Mick applied a tool from statistical process control, one whose usual application was in manufacturing. The tool was a *u*-chart, one of the many different kinds of control charts commonly used in SPC. Interestingly, something often overlooked when using control charts in real life, even by those well-versed in using the charts as guides for when to make changes in manufacturing or service processes, was that implicit statistical testing was going on in the charts. This point was discussed at length when Mick was deposed prior to the trial.

To set up the worth-1000-words graph considered in this section, we begin with the following table of sexual assault data. The counts in columns one, two and four were provided by DOC. Now, if the size of the inmate population were the same every year, then comparing the number of sexual assaults in one year to the number of such assaults in another year would have been fine. But since the population size changed every year, such a comparison could have been misleading. The appropriate things to compare were the annual rates at which such assaults occurred. Taking 5000 inmates housed for a full year as a convenient "unit" (hence the name *u*-chart), the professor was able to compare the number of sexual assaults per unit across years.

DOC Sexual Assault Data January 1, 1996 - September 30, 2006
Unit = 5000 inmates × years

Year	Sexual Assaults	Portion of Year	Average Daily Inmate Population	Units	Sexual Assaults per Unit
1996	3	1.00	18736	3.74720	0.8006
1997	4	1.00	20146	4.02920	0.9928
1998	4	1.00	20656	4.13120	0.9682
1999	3	1.00	20957	4.19140	0.7158
2000	7	1.00	20979	4.19580	1.6683
2001	6	1.00	20973	4.19460	1.4304
2002	12	1.00	21710	4.34200	2.7637
2003	85	1.00	22845	4.56900	18.6036
2004	167	1.00	23130	4.62600	36.1003
2005	178	1.00	22905	4.58100	38.8561
2006	170	.75	22897	3.43455	49.4970

The *u*-chart was one of many charts that the professor would explain when he taught a course on statistical process control, although the setting for his examples would have been manufacturing. Any good company that manufactured widgets would monitor its defect rate. But while a company might want to know the number of defective widgets per 5000 widgets produced each day, DOC might want to know the number of sexual assaults per 5000 inmates each year.

In the row for 2001 in the table, 20973 inmates equated to 4.19460 units [divide 20973 by 5000 to get this]. Since six sexual assaults were spread amongst 4.19460 units, that made 1.4304 sexual assaults per 5000 inmates [divide 6 by 4.19460 to get this]. It was 1.4304 and the other annual sexual assault rates found in the rightmost column that would be plotted on a *u*-chart in order to determine whether rates were holding steady or changing in some way. The *u*-chart, produced by Minitab® Statistical Software, is shown below.

The chart shows that prior to 2002, all points fell between two key borders (made up of horizontal line segments). There were standard procedures for how to use historical data to determine the location of these borders, called *control limits*. Someone familiar with SPC would say that sexual assault rates were stable (SPC jargon for *steady*) prior to 2002. By comparison, sexual assault rates for later years fell above the upper control limit, indicating that rates did not remain stable, i.e., that a possible change in the sexual assault rate had taken place. Not only was there an indication of an increase in the rate of sexual assaults in 2002, the alarm got louder every year thereafter.

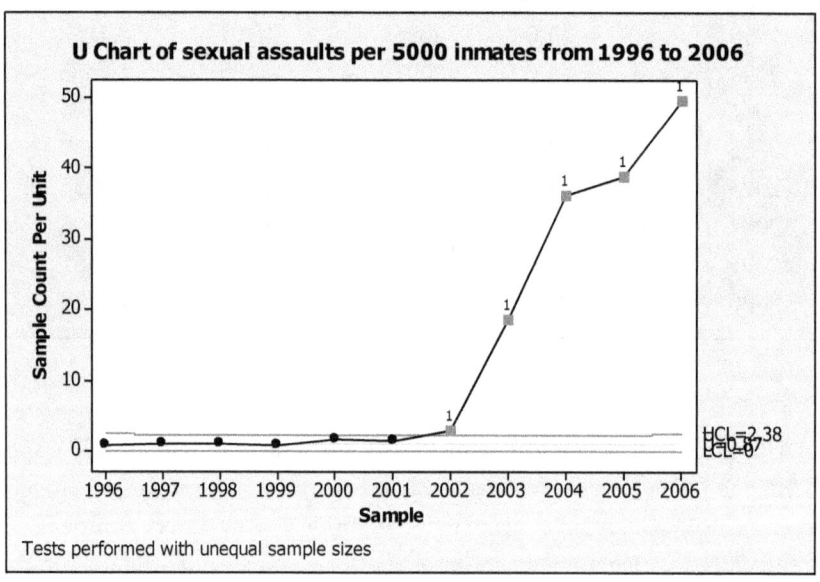

Behind the scenes, there was a lot of statistical theory associated with control charts. For example, seeing each year whether that year's point on the chart fell between the control limits was a statistical test. Also, the chart did not contradict an observation, made by an DOC staffer, that younger inmates were committing most of these infractions, that most older inmates wouldn't consider doing these things, and that the rise in successive years might be explained, at least in part, by younger inmates processing into the system as older inmates processed out. For sure, there was much statistical theory and interpretation behind the scenes, enough anyway that Mick was deposed for six hours by a plaintiff's attorney who was probing for weaknesses in the professor's report and for any weaknesses in how the professor might present himself in court.

The professor considered this case to be one of his more interesting expert witnessing gigs because it involved using a manufacturing tool in a non-traditional setting, it involved pink jumpsuits, and … well, because there was a picture worth 1000 words.

The reader might wonder what happened in the case. It ended when it became moot. The inmate who had brought the suit finished serving his time before the case was resolved.

35 REMEMBERING THE ANONYMOUS TEACHERS

It was a November morning. Technically, the date of his retirement had been October 1. Nevertheless, he would be teaching through to the end of the semester. There were not many empty cardboard boxes still left in his office. Every day, he'd been taking some books and personal items home. The idea that he would have to turn in the keys to the building and to his office at the end of the semester was really beginning to sink in.

The professor was already enjoying some of the benefits of retirement. Because it was past his formal retirement date, he was no longer a tenured member of the department, so he didn't attend tenure panel meetings. He also wasn't on any College-wide committees. And he had let go of worrying about which job applicants the department should make offers to. Of course, had he wanted to, he was welcome to go to recruitment meetings and offer input about whom to hire. But Mick felt it was better to have the people who had to live with the results of the hiring decisions make those decisions. Likewise, he didn't participate in committees that chose textbooks, since he didn't have a stake in which textbooks would be used in the spring.

Basically, all of activities that he still engaged in were ones he enjoyed. So during the last few weeks of his last semester, he felt a sense of release, much the same as if he were on sabbatical leave.

Checking his email that morning made him a smile. One was from Tracy. It had come in response to a work related email he had sent to former graduates of the department's masters degree program. Tracy knew that he was retired, but perhaps not that he was teaching through the entire the semester.

Dr. Norton

So good to hear from you. I'm glad that you're enjoying your retirement, though many students are missing out on the wisdom of a great teacher!! ☺ I use many things I learned from you in my classes.

Jennifer and I are both still teaching at the same high school.

I'll definitely pass the word along. We have some young teachers who might need the credits for recertification.

Continue to enjoy your retirement!

Tracy

So Mick was in a good mood as he turned toward his filing cabinet. It was almost empty. There were some very old teaching evaluations he could junk. As he skimmed them for nostalgia sake, he thought about how difficult it was to measure the quality of someone's teaching. This belief had been forged from the experience of having been on many tenure and promotion panels over the decades. He'd also, in behalf of colleagues, written numerous letters of recommendation in which he was to address a person's teaching ability. To gain insights about whether people were good teachers, he had observed many in action in the classroom. Mick believed that actually seeing a teacher in action was the most effective way to arrive at an opinion. Additional insights sometimes came from the personal experience of having taught someone's former students. He knew, for example, that Professor X had to be a good teacher because X's students were always well prepared for Mick's classes, whereas Professor Y's former students always seemed to be missing important skills.

Of course, the most convenient way to measure teaching was student evaluations. The information was there to be had, and it was numerical. The upside of being numerical was that it was easy to compare different people. The downside of being numerical was, well, …, gee. There actually were many downsides. The first to come to mind was that because anyone could compare two numbers – e.g., 5.3 is more than 5.0 – some peers looked to student evaluations almost to the exclusion of other evidence. Even a colleague in his own department often made observations such as So-and-so was below the departmental mean in four of the six Calculus I courses he taught during his probationary period. And Mick always felt silly pointing out that a person's evaluations tended to be consistent when teaching the same course repeatedly and that about half of the faculty in the math department would be below the mean, and be there consistently, and the other half of the department would be above the mean. That would be true even in a department that was full of exemplary teachers.

Over the years, he had examined many hundreds of student evaluations of other professors. Sure, he thought, teaching evaluations provided some insights in distinguishing truly outstanding teachers from truly poor ones. But he had seen some excellent teachers with a no-nonsense approach get poor evaluations, and some poor teachers with effervescent, likable personalities get great student evaluations.

One thing being overlooked by those who made too much of student evaluations was that in many courses, students couldn't know enough to say whether they were getting a good course. They could tell you if they were happy, but not much else.

He recalled an informal study he had done back when he was an untenured, greenhorn assistant professor. Back then, the teaching evaluation form that students filled out asked about a dozen questions. He was nonplussed to find that when a department chair or tenure panel examined student evaluations, item #6 – *Overall quality of instruction* was the only one that seemed to draw anybody's attention. It troubled him that the questionnaire had other items that ought also to have carried meaningful weight – e.g., *The instructor was prepared for class* and *The instructor was able to explain abstract ideas clearly*. When Mick expressed this concern to the chair of the math department, the latter's response had been, "You're in statistics. I'll get you some data and you tell me what you think."

The data Mick was given were from all of the sections of Calculus I and Precalculus taught that semester. To maintain confidentiality, faculty were identified as Instructor 1, Instructor 2, and so on. For each section taught, that section's mean was given for each of the items Mick said ought to be important. He was amazed at what he discovered. If an instructor had the 12th, say, best mean on item #6, he or she also had the 12th best mean on each of the other items that Mick thought were important. An instructor who was the 8th best on item #6 was also the 8th best on everything else. It was uncanny. Well, almost. Once in a rare while someone might be, say, 10th best on all but one item, and be 11th best or 9th best on that one.

The second amazing thing he noticed had to do with scale. Although each item had the same possible response scale (from 1 to 6), each item seemed to have its own empirical range. For example, the responses on item #6, an item on which students could respond viscerally, used up most of the range – if the students in a section loved the instructor, the mean might be 5.8; if the section hated the guy's guts, he might get 1.7. But even if the students hated the guy, they would admit that he was prepared, according the person a mean of, say, 5.1. But the 5.1 would be the lowest of all of the instructor means on that item. Mick didn't need to do a formal statistical test. It was obvious that the range of responses varied with the item, but an instructor's standing relative to his peers didn't. Since Mick refused to believe that the professor who was kth best in overall quality was also the kth best prepared, the kth best at explaining abstract ideas, and precisely the kth best at everything else, he concluded that all questions measured the same thing. He referred to it as *happiness level*. So he reported back to his chair that focusing just on #6 – *Overall quality of instruction*, was statistically defensible. He added that focusing on any one of those other items would have done just as well.

The professor sometimes mused about what the different responses would be like to a question posed to upper level college administrators nationwide – Rank the following from most desirable (1) to least desirable (4): happy students with good teachers, happy students with poor teachers, not-so-happy students with good teachers, not-so-happy students with poor teachers.

In cleaning out the filing cabinet, he finally came to the last item, a copy of newsletter that had been sent out to all of the faculty and students in the M. Ed. Program in Science and Mathematics. It was dated July 2, 2002 and contained the last Director's Corner piece he had written as director of the program.

Once upon a time (1968), I became certified to teach in the high schools of Missouri. Dr. Ken Stillwell, the man in charge of placing practice teachers, was very kind in placing me back in my hometown of Webster Groves, roughly 220 miles from Truman State University (known as Northeast Missouri State U. when I was a student there). This act had an unexpected consequence that changed my thinking entirely and forever on one particular subject. The cooperating teacher had taught me in the tenth grade. She had been, really, not one of my favorites. Actually, I considered her to be most unmemorable. Before gradually assuming all of her teaching duties, I needed to watch her teach for a few weeks. This was done from her desk, in the back of the classroom and by the door. I couldn't help but overhear some of the unkind comments students would make about her as they left class. Things hadn't changed much from when I was a student there. However, as I watched her teach so that I could phase into taking her classes, I watched her from a different perspective than I had six years earlier. What I saw was a true professional who used carefully chosen examples, spoke in carefully chosen words, addressed subtle details that had to come from her experience, and so on. She was a role model of a math teacher. All of this was beyond the notice of her students. It also was not lost on me that I had learned the material very well under her teaching. Somehow, it had never occurred to me to think about THAT before. I felt ashamed. Since watching my former teacher with different eyes, I have appreciated the professionalism of ALL teachers more, including the vast majority, the ones who don't get the accolades, the ones who just "do" their job.

36 GLIMMERS OF HAVING MADE A DIFFERENCE

Dr. Norton,

I was reading through the stats book that I use in my AP Statistics class and I noticed a problem titled, "Stand on Your Head to Lose Weight?" and thought, "Hmmmm.... something sounds familiar about that." I then realized that they wrote a problem about your research!

Anyway, I wanted to send you an email because I would love for you to come and talk to my AP Statistics class about your research and the things a "real statistician" does. My class meets from 2:30 - 4:00 in the afternoon on certain days of the week (it varies every week). Look and see if there may be some good days coming up where you could come and talk to them. I'd love for them to meet MY stats teacher!!

--

Kathryn Pedings
Math Department Chair
Charleston Charter School for Math and Science

ABOUT THE AUTHOR

Now retired from his day job at the College of Charleston, thereby having attained the lofty title *Professor Emeritus of Mathematics*, Mick Norton keeps busy by teaching occasional courses, working with students, and writing. And when gigs come his way, he continues with two of his hobbies that still remain – expert witnessing in statistics, and consulting on projects with manufacturers in the areas of statistics and statistical process control. Everything has its time, as did the favorite of all of his hobbies. The last wrestling meet he called was Davidson at The Citadel in 1998.

---

[i] Many of the incidents recounted herein are presented as conversations. Of course, no tape recorder would have been in use when an incident occurred. Because conversations presented are based on author's recollections, the author feels obligated to admit that his memory is imperfect. His other half assures him that this is so.

[ii] In real life, the faculty member who is described was hired after the expert witnessing case that is described in this section took place. To smooth the time travel that takes place in these sections – and sometimes to make a point – modest artistic license is taken with event chronology.

[iii] Citadel leads academies in sex assaults, *The Post and Courier*, Charleston SC, August 23, 2006.

[iv] *Cecilia*, written by Simon and Garfunkel, appeared in the album: *Bridge Over Troubled Water*, 1970.

[v] The names of some of the students, faculty and/or other identifying information given in these stories have been altered.

[vi] *Everything You Thought You Knew About Politics...And Why You're Wrong*. Kathleen Hall Jamieson, Basic Books, F Edition, 2000.

[vii] Exercise for statistics aficionados: Let d be the diameter of a bullet. Estimate the standard deviation of the horizontal distance between where the bullet lands and where it was aimed, given three hits in a row with the left-edge strategy followed by three in a row with the right-edge strategy. Make the simplifying assumption that the distance from rifle to target never changes. Also assume that the distance between where the bullet lands and where it was aimed is normally distributed.

[viii] Retrieved in 2003 (month and day somehow not recorded) from <http://powerliftingforum.net/forum/messageview.cfm?fID=1&tID=577.>

[ix] *Introduction to Statistics and Data Analysis*, Roxy Peck, Chris Olsen, Jay L. Devore, Cengage Learning, 2011, pp. 745-746.

[x] Senate panel cites "plague" of rural criminal violence. *The News and Courier*, Charleston, SC, June 19, 1991.

[xi] *When I'm 64*, written by Paul McCartney, appeared in The Beatles' Album: *Sgt. Pepper's Lonely Hearts Club Band*, 1967.

[xii] Actually, with the original data set no longer in existence, this data set is completely made up. No kidding. But it is made up so as to mesh with the discussion that is about to take place between James and Mick, which has a basis in fact.

[xiii] Portions of the input and output contained in this book are printed with permission of Minitab Inc. All material remains the exclusive property and copyright of Minitab Inc. All rights reserved. MINITAB® and all other trademarks and logos for the Company's products and services are the

exclusive property of Minitab Inc. All other marks referenced remain the property of their respective owners. See Minitab.com for more information.

xiv This dotplot was created using data – very sanitized – that appears in *A Quick Course in Statistical Process Control*, Mick Norton, Prentice Hall, 2005, page 111.

xv Retrieved October 5, 2009 from <http://www.charlestoncitypaper.com/charleston/opera--theatre-zwnj-52-pick-up/content?oid=1105214>.